Workforce Development for Global Manufacturing and Technology Transfer

Many countries are now emerging as new hubs for international manufacturing due to the increasing costs in traditional outsourcing locations. The success of this transition will rely heavily on the establishment and maintenance of a reliable manufacturing infrastructure through an integrated systems approach.

Workforce Development for Global Manufacturing and Technology Transfer: An Integrated Systems Approach focuses on work-integrated education, training, and learning. It offers practical case studies from the manufacturing and technology sector and advocates for the integration of people, technology, and processes to develop a critical workforce that aligns with national priorities. The book presents practical and modern strategies that nations, cities, and communities can adopt to develop and sustain manufacturing for sustainable industrialization.

While this book primarily focuses on manufacturing workforce development, it is relevant to manufacturers worldwide and applies to a global audience by covering topics such as project management, product development, technology justification, and various modes of technology transfer.

Systems Innovation Book Series

Series Editor: Adedeji Badiru

Systems Innovation refers to all aspects of developing and deploying new technology, methodology, techniques, and best practices in advancing industrial production and economic development. This entails such topics as product design and development, entrepreneurship, global trade, environmental consciousness, operations and logistics, introduction and management of technology, collaborative system design, and product commercialization. Industrial innovation suggests breaking away from the traditional approaches to industrial production. It encourages the marriage of systems science, management principles, and technology implementation. Particular focus will be the impact of modern technology on industrial development and industrialization approaches, particularly for developing economics. The series will also cover how emerging technologies and entrepreneurship are essential for economic development and society advancement.

Flexible Supply Chain: Industrial Engineering Optimization Modeling of the Postal Service System
Adedeji B. Badiru

Project Management for the Oil and Gas Industry: A Global System Approach, Second Edition, 2-Volume Set
Adedeji B. Badiru

Project Management for the Oil and Gas Industry: Systems Engineering and Technology, Volume 1
Adedeji B. Badiru

Project Management for the Oil and Gas Industry: Agile Systems and Environmental Concerns, Volume 2
Adedeji B. Badiru

Handbook of Construction Management: Scope, Schedule, and Cost Control, 2-Volume Set
Edited by Abdul Razzak Rumane

Construction Management: Quality Tools and Techniques
Edited by Abdul Razzak Rumane

Construction Management: Project Management, Process Principles, and Concepts
Edited by Abdul Razzak Rumane

Industrial Engineering Strategy for Constructive Technologies: A Systems-Based Approach for the Global Economy
Adedeji B. Badiru

Workforce Development for Global Manufacturing and Technology Transfer: An Integrated Systems Approach
Cassie B. Barlow, Sharon C. Bommer, and Adedeji B. Badiru

Workforce Development for Global Manufacturing and Technology Transfer

An Integrated Systems Approach

Cassie B. Barlow, Sharon C. Bommer, and Adedeji B. Badiru

CRC Press
Taylor & Francis Group
Boca Raton London New York

CRC Press is an imprint of the
Taylor & Francis Group, an **informa** business

Designed cover image: Shutterstock

First edition published 2026
by CRC Press
2385 NW Executive Center Drive, Suite 320, Boca Raton FL 33431

and by CRC Press
4 Park Square, Milton Park, Abingdon, Oxon, OX14 4RN

CRC Press is an imprint of Taylor & Francis Group, LLC

ISBN: 978-1-032-88362-5 (hbk)
ISBN: 978-1-032-88435-6 (pbk)
ISBN: 978-1-003-53775-5 (ebk)

DOI: 10.1201/9781003537755

Typeset in Times
by KnowledgeWorks Global Ltd.

To those who labor to contribute to manufacturing and the advancement of economic well-being in our society.

Contents

Foreword

In this book, authors Barlow, Bommer, and Badiru present a diverse set of thought-provoking perspectives and prescriptions concerning the challenges and other factors influencing workforce development in the manufacturing sector – an industry environment characterized by increasing globalization and rapid technological change. While the authors' primary focus is on workforce development in the manufacturing sector, many of their observations, arguments, and recommendations are readily generalizable to other sectors of the economy, e.g., workforces in the services sector, government, R&D, academia, etc.

I can relate. Over my 50+ years working full time, I gained a considerable amount of workforce development experience, as both an educator and a practitioner. As an educator, I helped prepare both practicing engineering professionals and technicians to more effectively meet the current and future challenges in their respective jobs. As a practitioner, I was privileged to have had the opportunity to lead both small and large organizations. In that role, one of my primary responsibilities was to recruit and professionally develop the workforce members – both as individuals and as a team – to continually improve our organization's effectiveness, efficiency, resilience, and sustainability, and to meet the challenges we were likely to face in the future. So, I can personally relate to – and value – many of the observations and recommendations presented by the authors.

A point of emphasis. From a total system perspective, the manufacturing workforce is more than the equipment operators, assemblers, craftsmen, and others involved with actually making products – many more people. The total workforce involves everyone involved – directly and indirectly – in producing the product, in desired quantities, quality, cost, and timeliness. For example, the workforce also includes product designers, engineers, logisticians, quality control specialists, marketing and sales people, financial analysts, legal staff, human resources staff, managers, and many others. Workforce development includes recruiting the right numbers of people, with the requisite skills, talents, and expertise and then continuously improving the effectiveness and efficiency of all of these people, as individuals and especially as teams, through continuing education, training, varying job assignments and content, etc.

In many cases, it is more cost-effective and efficient to outsource a part of the production process, rather than establish, develop, and maintain that capability organically. The workforces of these partners (e.g., subcontractors) are a de facto part of the total system workforce required to produce a product (again, in the required quantities, quality, desired price point, and schedule). In my own experience, albeit in a service (non-manufacturing) situation, I found it necessary to provide our partners' workforce with training and oversight necessary to accomplish our desired outcomes.

Today's work environment is, in part, characterized by increased globalization, in which an organization must consider foreign suppliers, competitors, international logistics, and market opportunities. For some organizations, that also means establishing operating/manufacturing locations in foreign countries, employing foreign nationals, and operating in respect of local laws, regulations, customs, and social norms/expectations. In these situations, workforce development can vary significantly from those tactics employed with workforces in the US. In my own case, I had the opportunity to lead an organization based in Germany, in which half of the 750-person workforce were German personnel, while the other half were US citizens. The German part of our team had to be led and administered in compliance with German labor laws and regulations, as well as German cultural and social norms.

Perhaps the biggest challenge to workforce development these days has been, and will continue to be, the effects of rapid and pervasive technology change, both evolutionary change and revolutionary change. Not only is technology changing in a general sense, but in many sectors, it's changing at an accelerating rate. This often requires more rapid adaptation and more rapid decision-making by the organization's leaders, often under increased uncertainty and associated risk. As new technologies are adopted in the manufacturing sector of industry, the workforce needs to (often rapidly) adapt to the changing demands of these new technologies. This is, and will be, a continuing challenge for workforce development. When I initially trained as an engineer (too) many years ago, the technology of the day involved slide rules, blueprints, and simple calculators. Those days are long gone. Today's recent college graduates have never known a world without (e.g.) the internet, personal computers, wireless communication, the Global Positioning System, computer simulation, additive manufacturing, or (sadly) cybersecurity issues. Those technologies were largely developed during their lifetime – in just a single generation. The developing technology of Artificial Intelligence ("AI") will undoubtedly have a significant impact on all sectors of industry and, most likely, on society in general. So what? In today's environment, to remain competitive, managers and supervisors must continually assess the implications of adopting new technologies on the workforce, e.g., the numbers of people required, the skills and talents they will need to perform effectively and efficiently with the new technology, and the continuing education and training they will need to remain competent in the future.

The authors emphasize the utility of taking a "systems approach" to workforce development. I completely agree. This is often characterized by a process model that sequences INPUTS > TRANSFORMS > OUTPUTS. INPUTS to the system process come in two general "flavors." "Resources" are those inputs over which the organization has some control or about which it can make decisions. These include (e.g.) people (the workforce), raw materials, process technologies/equipment, price points of products to be manufactured, financing to be acquired, and so on. In contrast to resources, system INPUTS also include "environmental factors" which affect the system process, but over which the organization has no direct control, e.g., laws, government regulations, competitors, customer demand, societal norms, etc. In mathematical programming speak, resources are decision variables, while the environmental factors are constraints. System TRANSFORMS are those processes that organize and coordinate the inputs in some fashion to produce the desired OUTPUTS, e.g., the products being manufactured at some planned quantity, quality, cost, schedule, etc. But what is really important are the system process

OUTCOMES. Do the system OUTPUTS actually achieve desired goals of (e.g.) satisfying customer demands, revenue, profit, brand recognition and respect, etc. Environmental factors are variable – the actions of competitors, suppliers, government legislators and regulators (and others) are not constant and combine with system OUTPUTS to determine actual OUTCOMES. These changes in environmental conditions are often unpredictable and unanticipated. This requires the system to provide continuing assessment and feedback to decision makers and, as necessary, adjustments to the process (i.e., adjustments to the inputs and/or transform processes/technologies) to address deviations of actual outcomes from the desired/planned outcomes. This is a "cybernetic" model involving both responsive feedback and feed-forward channels. One of the adjustments that should always be considered is any necessary changes to the workforce, its composition, organization, skills and talents, etc. How should it be developed going forward?

Just as the workforce is an important component of the manufacturing system, I contend the workforce itself should be viewed – and developed – as a system. The authors rightly observe that a system is a set of components which, operating in relation to one another in some fashion, produce outcomes that are greater than can be achieved by the components operating independently ("The whole is greater than the sum of the parts"). While it is certainly important to develop the skills and talents of the individuals in the workforce, in my experience, it is even more important to develop the workforce *as a team.*

I hope you will find this text as thought-provoking as I did. It examines and underscores the important implications of globalization and technological change on the continuing development of the (total) manufacturing workforce. In the final analysis, organizational success in manufacturing and in other segments of the economy is *all about the people.*

Todd I Stewart
Major General, USAF (Ret.)
Chancellor, Air Force Institute of Technology (Retired)

Preface

Global interfaces are essential for national advancement and sustainability of economic vitality. Manufacturing is the primary foundation for global interfaces that provide benefits to all parties involved. Global manufacturing technology transfer presents strategies, adaptation methods, and management techniques for the commerce and trade connections of the USA to other countries. The premise of this book is to advance existing manufacturing infrastructure through workforce development, using an integrated systems approach.

Africa and many less-glamorous countries are emerging as new hubs for international manufacturing as a result of rising costs in traditional outsource destinations in Asia. Many manufacturers are returning previously outsourced manufacturing operations to their home countries, but they are still interested in exploring new outsourcing options. As a result of these new developments, Africa and many parts of Europe are poised to assume a leading role in global manufacturing. However, the foundation for doing so successfully will depend on the development and sustainability of a reliable manufacturing infrastructure, using an integrated systems approach. A key requirement for this will be new and innovative mechanisms for bi-lateral technology transfer. Although this book focuses specifically on manufacturing workforce development, it is equally of direct interest to all manufacturers in diverse parts of the world, particularly in Europe. This book's contents are applicable to different audiences globally. Contents include project management, product development, technology justification, and technology transfer modes. The entire world is a global system of systems. We are all interconnected through one medium or another. Global manufacturing is, indeed, one of the influential systems within the world system. The way we are intertwined and interconnected around the world, we must understand our interdependency. This book, *Global Manufacturing Technology Transfer*, presents practical modern strategies that nations, cities, and communities can adopt for developing and sustaining manufacturing for the purpose of sustainable industrialization. The economic development and industrialization of any nation depend on solid foundations of manufacturing. Over the passage of time, industries come and go and requirements for economic commerce shift. Nations that survive and continue to thrive use adaptive strategies that meet current and future needs. If good sustainable strategies are available, organizations and communities can survive any turbulence that develops in the normal course of economic development. This book is operationally focused, with adaptive systems content to fit the needs of different manufacturing global audiences. This book uses case studies from the USA Midwest to demonstrate how modern technology innovation strategies can be developed and harmonized to keep a nation moving forward economically. USA manufacturing technology transfer models and techniques are applicable to other technology transfer relationships between any pair of collaborators. Although Africa is a continent, and the USA is a country, trade and commerce between the two entities are similar to any country-to-country relationships. Indeed, within the Africa-USA relationship reside country-by-country interfaces of the USA with specific countries of Africa and

Europe. These international relationships can be reformulated to facilitate workforce development in a constructive manner, whereby workforce development needs of the USA can be filled by constructive inputs from partnering countries. Some examples include USA-Nigeria, USA-Liberia, USA-South Africa, USA-Sudan, USA-Ghana, USA-Egypt bi-lateral relationships, and so on. The Sub-Saharan countries of Africa are more noticeable when the issues of developing and underdeveloped countries are covered in literature or popular press. The economic size of the USA does, indeed, put it on a comparative par with the continent of Africa. There are several global issues affecting how nations interact politically, socially, militarily, economically, and commercially.

Acknowledgments

We thank all our colleagues, partners, collaborators, and supporters in our respective organizations over the years: SOCHE, University of Dayton, and Air Force Institute of Technology. We continue to be inspired and motivated by discussions and ideas raised by colleagues, leading to the motivation to write this book. Although too many to name individually, we, particularly, appreciate the help of Ms Lisa Smittle, Dr Betsy Grimes, Dr Heidi Ries, and other AFIT colleagues. We would also like to thank Ms Michaela Finn for providing an initial cursory edit of this book manuscript.

About the Authors

Dr Cassie B. Barlow is the President of the Strategic Ohio Council for Higher Education (SOCHE). Through SOCHE's multiple programs, Dr Barlow is focused on collaboration among SOCHE's 23 College and University members as well as professional development of faculty and staff and workforce needs of the almost 200,000 students at the regional institutions. Dr Barlow is familiar with leading multiple activities. She was the 88th Air Base Wing and Installation Commander, Wright-Patterson Air Force Base (AFB), OH. In that role, she commanded one of the largest air base wings in the Air Force with more than 5,000 Air Force military, civilian, and contractor employees. Dr Barlow retired from active duty with the US Air Force as a Colonel in 2014. Dr Barlow was commissioned in 1988 as a distinguished graduate of Georgetown University, Washington, DC. Following graduation, she was assigned to her first of 13 assignments spanning the world. Dr Barlow earned a Bachelor of Psychology degree from Georgetown University, Washington, DC.; a Master of Clinical Psychology degree from Chapman University, Orange County, CA; and a PhD from Rice University, Houston, TX. She also earned a master's degrees in Military Operational Art and Science, Air Command and Staff College, AL, and in National Security, Industrial College of the Armed Forces, Washington, DC.

Dr Sharon C. Bommer is Associate Professor in the Department of Engineering Management, Systems, and Technology at the University of Dayton. She holds a PhD in Engineering, specializing in Industrial and Human Systems, with a focus on human performance and cognition. Dr Bommer's educational background includes a Bachelor of Science in Mechanical Engineering, along with two master's degrees: one in Business Administration and another in Industrial Engineering. Before completing her PhD, Dr Bommer amassed over 15 years of experience in the automotive manufacturing industry. She held positions at General Motors and Toyota, where she worked in various engineering and operational roles. Before pursuing her doctoral degree, her final position was as a Plant Manager for a Tier 1 parts supplier. Dr Bommer's career in academia includes serving as the Dean of the School of Business and Applied Technologies at Clark State College. In this role, she provided leadership for programs in business, information technology, agriculture, and engineering technology. She has a passion for teaching both graduate and undergraduate courses, particularly those focused on operational improvements. Her research primarily revolves around human systems integration, with a specific emphasis on human performance and cognition in industrial and military operations. Dr Bommer has contributed to the field through her publications. She co-authored the technical reference book *Work Design: A Systematic Approach*, published by Taylor and Francis CRC Press, and authored a chapter in the prestigious *Maynard Industrial Engineering Handbook 6th Edition*. Additionally, she has contributed to various peer-reviewed journal articles, conference proceedings, and technical presentations.

Dr Adedeji B. Badiru is Dean Emeritus, Graduate School of Engineering and Management at the US Air Force Institute of Technology (AFIT). While serving as Dean, he had oversight for planning, directing, and controlling operations related to granting doctoral and master's degrees for the US Air Force. He was previously Professor and Head of Systems Engineering and Management at AFIT, Professor and Department Head of Industrial Engineering at the University of Tennessee, and Professor of Industrial Engineering and Dean of the University College at the University of Oklahoma. He is a registered Professional Engineer (PE), a certified Project Management Professional (PMP), a Fellow of the Institute of Industrial & Systems Engineers (IISE), a Fellow of the Industrial Engineering and Operations Management Society (IEOM), a Fellow of Institute for Operations Research and Management Science (IORMS), and a Fellow of the Nigerian Academy of Engineering (NAE). He is also a Program Evaluator for ABET. He holds a leadership certificate from the University of Tennessee Leadership Institute. He has a BS in Industrial Engineering, an MS in Mathematics, and an MS in Industrial Engineering from Tennessee Technological University, and a PhD in Industrial Engineering from the University of Central Florida. His areas of interest include mathematical modeling, systems design and integration, project management and optimization, innovation management, and productivity improvement. He is a prolific author, with over 45 books, over 40 book chapters, over 200 journal and magazine articles, and over 220 conference presentations. He is a member of several professional associations and scholastic honor societies. Professor Badiru, a world-renowned educator, has won several awards for his teaching, research, administrative, and professional accomplishments. He is the 2020 recipient of the Lifetime Achievement Award from Taylor and Francis Publishing Group. He was also a member of the AFIT team that won the 2019/2020 US Air Force Organizational Excellence Award. Dr. Badiru is also the recipient of the 2022 BEYA Career Achievement Award in the Government category and the 2024 Donald G. Newnan National Engineering Economy Teaching Excellence Award from the American Society for Engineering Education (ASEE). He also received the 2024 Tau Beta Pi Distinguished Alumnus award. In 2009, he received the Wellington Award from the Engineering Economy Division of the Institute of Industrial and Systems Engineers. In 2022, Dr Badiru received the Frederick Taylor Award, the highest award from IEOM (Industrial Engineering and Operations Management Society). He holds a US Trademark for DEJI Systems Model for Design, Evaluation, Justification, and Integration. Badiru is a member of Fulbright Specialists Program.

Affirmation Statement

Creatively adjusting to address quickly evolving workforce training requirements is key to global competitiveness. As digital engineering, digital twin, thread requirements, advanced manufacturing processes, artificial intelligence adoption, and industrial internet of things capabilities advance and transition from optional points of interest to necessities, it is incumbent on those in academia, industry, and government to collaborate to ensure an integrated approach. The future promises yet-unimagined advances and benefits, if these new technologies and innovations can be harnessed and embedded at all levels of workforce development efforts.

Dr Andrew D. Shepherd
Chief Research Officer
Sinclair Community College

1 Labor in the Workforce

INTRODUCTION TO THE VALUE OF LABOR

You may delay, but time will not.

Benjamin Franklin

The issue of labor is pervasive and urgent all over the world. Timely attention is essential. Labor is the backbone of the workforce in any organization, particularly in manufacturing and technology enterprises (Badiru, 2019; Badiru et al, 2019). The Industrial Revolution, which defined the turning point in the advancement of our society, happened on the back of labor, aided by radical mechanization (Badiru, 1993). Labor, in the context of this book, entails both the physical (brawn) and intellectual (brain) assets of an organization. Labor must be taken into account in any workforce development program. The recurring clamoring for new labor unions in business and industry is indicative of the need to pay more attention to workers' hierarchy of needs (Maslow, 1943). This book, written by three long-term workforce advocates, aims to spark appropriate interest and renewed vigor in workforce development programs.

Elon Musk is reputed for saying:

If you don't make stuff, there is no stuff.

Elon Musk

Labor (the nucleus of the workforce) makes stuff for the benefit of humankind. Consequently, developing, retaining, preserving, and sustaining the workforce have direct implications for economic and social well-being of our society. For this reason, the premise of this book is workforce development, specifically, for manufacturing enterprises. The goal is to prepare the workforce now for the nature of work for the future. Aligning the workforce to the digitally driven workplace of the future requires a systems-thinking approach. Such is the methodology of this book. We take a deep and diverse look at all the factors that can impinge upon the preparation of a versatile, ready, and resilient workforce. This includes a consideration of what constitutes labor, which is the foundation for human-resource constituents of the workforce. Clark et al (2008) highlight the fact that working people built America. Labor innovations, investments, and shrewd management by the titans of industry at the turn of the century in America laid the foundations for many subsequent industrial development results. The works of John Piermont "JP" Morgan, Henry Ford, John D. Rockefeller, Andrew Carnegie, and Cornelius Vanderbilt have continued to reverberate through modern industries around the world, although the moniker of "The Robber Barons" is sometimes used, unjustifiably by detractors to describe the

DOI: 10.1201/9781003537755-1

fivesome (quintet). The history of the American Gilded Age (1877 to early 1900s), known for economic growth, technological advancements, and sustainable industrialization, cannot be separated from the history of labor advancements. The importance of value-adding labor must be recognized and celebrated.

COST OF LABOR AND THE ECONOMIES OF SCALE

Computationally, the work rate of labor has been shown to have a direct correlation to the economies of scale (Badiru & Omitaomu, 2007), which has facilitated the expansion of the industrial base. Physical reduction of labor, as a group, does not, necessarily, translate to production cost savings unless an appreciable increase in the individual work rate could be achieved. Several human-based factors, personal or organizational, can influence the total work rate and production output of the workforce.

Savchenko (1990) provided a treatise on the importance of labor in the context of social, political, and economic development of a nation.

"What is Labor?" holds a special place in the series ABC of Social and Political Knowledge. This book deals with the role of labor in the rise of man as the main productive force and a vehicle of production relations.

Labor, its conditions, and results are studied by different sciences. Natural sciences study labor in its psycho-physiological aspects, while social sciences examine it as a socio-economic phenomenon. Each science uses its own approaches and methods.

Marxist political economy ranks first among the social sciences studying labor. It deals with the social form of labor and the social structure of labor, that is, relation among people stemming from their participation in social labor.

The author describes the historical stages in the development of labor from the point of view of both their common and specific features. Labor has always been a domain of exchange between man and nature. The content of labor may remain the same at different stages in man's history. For instance, the transition from feudalism to capitalism is not necessarily accompanied from the outset by changes in the technical basis of production. The character of labor, however, undergoes revolutionary changes whenever one mode of production is superseded by another.

Objectively essential to man's vital activity, labor is his eternal companion. However, the essence of man's attitude to labor differs depending on the social form of the latter. The author focuses on the analysis of the social form of labor under capitalism and socialism.

Under capitalism, labor takes on the form of hired labor. Hired labor is a historical form of labor which emerges when the means of production become counterposed to man as an alien force, as capital. Man cannot work before he sells his labor power, i.e. his ability to work, to the owner of capital – the capitalist.

Their interests can never coincide because the capitalist uses the means of production, machinery, and technology, which are in possession, to exploit hired workers. For all variety of its forms, coercion is the common feature characterizing labor in antagonistic societies where the coercive nature of labor is predetermined by its social form.

Socialism has eliminated the hired labor system and modified the mode in which the means of production and labor power come into contact. With exploitation and economic coercion to work ruled out, the character of labor fundamentally changes. Under socialism, it is the public duty of all members of society to work in order to satisfy their needs. This duty is a law of life. Whereas under capitalism scientific and technological progress conduces to growing unemployment, under socialism, it ensures full employ employment.

Under socialism, man's labor is a measure of his participation in social production and a criterion relied on in distribution and consumption. The higher the productivity and effectiveness of social labor, the greater the opportunity to raise the level of well-being of society as a whole and of each of its members. However, under socialism, labor is not yet man's prime vital need. It is primarily a means of subsistence. At the same time, under developed socialism, there emerges an economic basis for labor to increasingly assert itself both as a means of satisfying social needs and as a prerequisite for the development of man's abilities.

Labor is a most important factor in the evolution of world civilization. However, it is only under socialism that it turns into the main stimulus for the development of social humanism and optimism.

DIGNITY OF LABOR IN THE SOCIETY

Man, just as the society he lives in, is inseparable from nature with which he constantly interacts. Man cannot exist without providing for his primordial, natural, biological needs – the need for food, clothing, shelter, protection against the harmful effects of the elements (natural calamities), etc. Nature cannot provide everything that man needs in a ready-made form.

Much has to be produced by man himself. That means that man has to work. As he starts producing the goods he needs, man comes into direct contact with nature. Even the gathering of fruit that nature offers to man in a ready-made form, even hunting and fishing, constitutes a process of interchange between man and nature, a process through which man exerts his influence on nature and transforms natural conditions.

As labor processes developed, man distanced himself from nature. Engels provided a brilliant analysis of the historical significance of labor in his article "The Part Played by Labor in the Transition from Ape to Man" (1876). He examined the rise of man and primitive human association in the course of the development of human labor, labor activity, and production by man of the means and conditions for his own survival. Thus, labor itself is among man's basic vital needs. The genesis of human society is inseparably linked to man's labor activity and to the development of material production.

What Is Labor?

Labor is man's purposeful activity through which he adapts natural objects and uses them to satisfy his needs. In any labor process, man expends his physical, nervous, and mental energy. Labor results in the emergence of useful products.

However, the role of labor in man's life is not limited to its serving as a necessary condition of man's existence and development, to its being a source of his strength and wealth. While acting on nature in the process of labor, man transforms himself, develops his materials and spiritual culture, his physical and mental abilities.

Historical Perspective

At first, man obtained the material wealth he needed from nature in a ready-made form. This stage lasted for many hundreds of thousands of years. As time went by, man learned to make things from the available primitive natural materials such as wood and stone. In order to secure his vital needs, man had to constantly improve his tools and process of making them. As he did so, his labor skills were perfected.

It is due solely to labor that the human hands have attained a degree of perfection high enough to conjure into being irrigation systems, the Eiffel Tower, the statues of Michelangelo, and color television. In their collective labor, human beings arrived at a point where they had something to say to each other. Necessity conditioned the emergence and development of articulate speech. As human beings interacted and communicated in the process of labor, their mental abilities developed. Today, man's mental abilities have reached a truly fantastic level making possible the creation of spacecraft, computers, nuclear reactors-in a word, making it possible for man to translate his most daring dreams into life.

At first sight, man's labor resembles the "work" of animal, for instance, ants, bees, beavers, and birds. However, man's labor and animals' "work" are essentially different. While human labor is always a purposeful, rational activity, animals are guided by instincts. The worst architect differs from the best bee in that before building a honeycomb he has it built in his mind.

Another characteristic feature of man's labor lies in his using implements of labor he has made himself to produce various goods. Some animals are known to apply "tools" in procuring food. The Egyptian vulture, for instance, uses stones as tools. The Galapagos' woodpecker uses little pointed sticks to pick insects out of wood. The sea otter may use a stone to pry mollusk shells open. The chimpanzee applies a whole variety of tools. However, these examples illustrate something that is occasional in the life of the animals. None of the animals makes implements of labor. Through his labor man harnesses the forces of nature making them serve his goals, while animals adapt themselves to these forces without being able to consciously act on them or master them. In this lies man's essential difference from the animal, a feature he owes to labor.

As the educational functional of labor evolves, labor turns into a means of self-expression, a means of developing man's creative potential.

The need for labor rests on a natural foundation since it is dictated by human nature.

Thus, there are three important elements in the process of labor.

The first one is labor per se, or labor as a purposeful activity.

The second essential element is the implements of labor, sometimes referred to as means of labor – the things man uses to act upon nature and to adapt natural objects for his own use. There are material and mental means of labor. Among them, a special place is held by the means of production – tools, machines, instruments, etc.

Man uses the means of production to reinforce, as it were, the organs of his own body. For instance, unable to discern the structure of an animal or vegetable cell with his naked eye, he has equipped himself with an electron microscope.

Production premises, structures, installations, ventilation systems, railways, highways, other roads, and canals are also important means of labor. Land, too, is a means of labor as it is the place of labor, an embodiment of natural properties essential to agricultural production and a storeroom of natural wealth (iron ore, diamonds, gold, bauxites, copper, uranium ore, oil, gas, potassium salts, etc.).

Many millennia ago, man first took a stick or a stone-in his hand, thus enhancing his natural strength. Today, using the tools he has made, man has succeeded in "reaching" the Moon. The Mars and the Venus.

The entire history of man's conquest of nature is contained between these two extreme points of progress. It is, however, only in a figurative sense that one can talk about man conquering nature. Man is a part of nature. The man's life – labor – is inseparable from the material provided by nature. But the point is that unveiling the secrets of nature and opening its treasure-houses calls for continuous and intensive labor. The efficiency of this labor is the higher, the greater the number of electric, mechanical and other appliances, or "helpmates" that man employs.

Here are some data that give a general idea of the role of the "helpmates" man relies on the process of labor. Over a century ago, human manual labor accounted for 15 percent of the energy expended in the production of goods (draught animals and machines accounted for 79 and 6 percent, respectively).

At present, the share of human labor is 3 percent, while the proportion of the energy expended by draught animals and machinery is 1 and 96 percent, respectively.

The third element in man's purposeful activity is the object of labor, i.e. the materials which are to be transformed or processed. Some of the objects of labor – for instance, woods, coal- and oil-fields, ore and diamond deposits, etc. – are provided by nature. These are primary objects of labor. The rest have to undergo preliminary treatment before further use. For instance, grain has to be milled into flour before baking, and coke has to be fired before it is used for smelting. These objects of labor are referred to as raw materials or secondary objects of labor.

While primary objects of labor predominated at the dawn of social production, secondary objects of labor acquired importance at subsequent stages of social production. In our time and day, man-made objects of labor – artificial fibers, plastics, polymers, alloys, etc. – have a major role to play. There are also supplementary materials used in production – fuel, lubricants, etc.

The instruments and objects of labor essential to manufacturing the things that man needs are the material elements of the labor process.

In their totality, they constitute the means of production.

Although the means of production have a major role to play in the process of production, they would be useless without man. Yarn, unless it is used for knitting or waving, is just cotton or wool put to waste. Leather would never turn into footwear unless man applies his labor to it. For yarn, leather and like to become useful, "live" labor of conscious human beings must be applied to them. To this end, the means of production have to be brought into contact with man's physical and mental abilities, occupational skills, and production experience, which he uses to produce material goods with the help of the means of production.

The means of production and man's labor power constitute in their totality the productive forces of society. The growth and improvement of the productive forces give man more power over the forces of nature and lay the foundation for the development of human society.

The dominant role in the productive forces belongs to people. Not only do they use the means of production, they make them. Endowed with an ability to work and to set in motion the means of production, people constitute the main element of productive forces.

Man has never produced material goods alone. Even primitive men hunted and tilled land together. Modern enterprises sometimes employ many thousands of workers. Each enterprise receives raw and other materials from dozens and sometimes hundreds of enterprises. For Instance, automobile factories rely on hundreds of suppliers. Knowledge, skill, and experience are used on an increasing scale. Thus, in the process of labor human beings come into contact with one another of necessity, which makes their labor social by nature. Production is always social production and labor is always social labor.

Those who argue that labor is not necessarily social cite the example of Robinson Crusoe, a character invented by the English writer Daniel Defoe. During his 30 years on a desert island, Robinson Crusoe engaged in farming, cattle breeding, and handicrafts. However, people tend to forget that he had at his disposal the instruments and tools he had obtained from his wrecked ship and that he grew crops from the grains he found in his pockets. The tools and the seeds were the results of the labor put in by many other people. Moreover, Robinson Crusoe relied on the skills he had acquired from other people. And yet, Daniel Defoe could not make his hero live in complete isolation for a long time, so he let Friday keep Robinson Crusoe company. Thus, Robinson Crusoe's story does not disprove the argument that human beings have always procured the means of subsistence jointly rather than in isolation.

In the process of production, distribution, exchange, and consumption of material goods, people inevitably, involuntarily, and unconsciously enter into social relations, which are referred to as production relations. Man's labor can take place and production can be maintained only within the framework of these social ties and relations. These relations wholly depend on the form of ownership of the means of production.

A form of ownership is a certain mode in which people appropriate the means of production and, consequently, the fruit of other people's labor.

The position of an individual in the system of production relations depends on who owns the means of production.

The position of the owner of land or any other private property – factories, workshops, banks, etc. – radically differs from that of the worker, engineer, technician, or anyone else who sells his labor power to those who own the means of production. People's relations with respect to the ownership of the means of production thereby predetermine the objective difference in their social position, their relation to one another in the process of labor, and their place in the social system as a whole.

Private ownership of the means of production predetermines the lasting dependence (both economic and extra-economic) of the majority of participants in the material social production process, those who are deprived of the means of production, on an insignificant minority who own the natural wealth, workshops, factories, retail shops, banks, and other means of production.

As human society developed, different forms of private ownership – slave-holding, feudal, and capitalist – emerged.

In contrast to the forms of ownership listed above, public ownership of the means of production is based on the collective appropriation by the people of the means of production. Hence, all the participants in the production process, i.e. all the members of society, have the same relation to the material conditions of their labor activity. Distinguished among the forms of ownership are group ownership, which existed in primitive-communal society, public (state) socialist ownership (whereby the means of production are the property of the entire people), and collective-farm–cooperative socialist ownership (whereby the means of production belong to work collectives).

The highest form of public ownership will emerge in the future as communist ownership of the means of production.

Historically, each of these forms of ownership has manifested itself in a specific mode of production, i.e. a way whereby the producers come into contact with the means of production.

Primitive-communal production relations were characterized by the communal (group) ownership of the primitive means of production.

In slave-holding societies, production relations were specific in that the means of labor and the worker (the slave) belonged to the same owner.

Both the means of labor and slaves, being the property of slave-owners, were treated as things.

Feudal production relations, based on large-scale land proprietorship (landlordism), were highlighted by the exploitation of the peasants by the landowner (the feudal lord). The peasants were forced either to farm the feudal lord's land without pay or to regularly pay him the rent and duties.

Capitalist production relations are based on private ownership of the main means of production (factories, mines, pits, etc.). Under capitalism, the producers, i.e. the labor force, are free from personal bondage to the owner of the means of production. Personally free, but deprived of the means of production, the producers are forced to sell their labor power in order to secure the means of subsistence for themselves. Having acquired their labor power as a commodity, the owner of the means of production (the capitalist) seeks to make a maximum use of it in an effort to amass profit. Thus, on the one hand, the producer gains access to the means of labor and, by coming into contact with them becomes involved in the process of production. On the other hand, the employer sizes the opportunity of getting richer by appropriating the surplus value (profit) produced by the labor force over and above its own value. The profit is appropriated by the capitalist.

Things are radically different under socialism. Socialism constitutes the first stage of communist production relations and features public ownership of the means of production. The latter eliminates the antagonistic character of production relations. In laying foundation for a fundamentally new type of production relations, socialism puts an end to the exploitation of man by man and eliminates conditions essential to the existence of those social classes which live by appropriating the results of other people's labor.

Nonetheless, with the establishment of socialist economic relations, the differences between the two main working classes – the workers and the peasantry – do not immediately disappear altogether. They persist because socialism does not immediately remove the differences between the two different types of social property – state and cooperative-collective-farm – in the means of production. These differences will be overcome as developed socialism evolves into communism.

Thus, the history of human society shows that the development of production, i.e. of each concrete mode of production in the totality of productive forces and production relations involved in it – is determined by the form of property. The passage from the primitive-communal to the slave-holding society and then to the feudal, capitalist, and communist societies is the history of the different modes of production superseding one another. Basically, it is the history of the successive forms of ownership of the means of production which likewise supersede one another.

Why does one mode of production give way to another mode of production?

Production relations are brought into existence by definite productive forces and, for their part, they exert a strong influence on the latter.

Progressive production relations, which correspond to the nature of the productive forces, promote and accelerate the latter's development and stimulate material production as a whole, enhancing its effectiveness. Outdated production relations, on the contrary, hinder and slow down the development of production forces.

This happens according to the law of correspondence of production relations to the character and development level of productive forces. As a mode of production based on private ownership and exploitation develops, this law is violated. The incongruity is caused by the contradiction between the relatively mobile productive forces and the production relations that tend to lag behind and slow down the former's development. As the contradiction deepens, there arises a need for a radical change in the social organization in whose womb this contradiction gradually develops into a conflict. There emerges a need for a social revolution to remove the outdated production relations and institute new ones.

Production relations, i.e. relations of ownership, production, exchange, distribution, and consumption in their aggregate, constitute the basis or the economic structure of a society. It is on this basis that various social relations, ideas, and institutions appear and develop. The basis of each society underlies the latter's superstructure, i.e. political, ideological, legal, moral, and religious relations backed up by corresponding organizations and institutions (the state, political parties, social organization, the church, etc.) and forms of social consciousness (ideas, theories, views). The superstructure is class-oriented as it serves and protects the basis, expressing the interests of the society's predominant class. In the course of a social revolution, the old basis is eliminated and the superstructure undergoes a simultaneous change.

In a society that has attained a certain stage in its historical development, the mode of production, the economic structure (the basis), and the corresponding superstructure are inseparably linked to one another, constituting an integral organism referred to as a socio-economic formation.

The historical process is a succession of socio-economic formations which supersede one another. Each successive formation constitutes a higher scale of civilization, a higher scale of civilization, and a higher level of social development.

In class-antagonistic socio-economic formation (slave-holding, feudal, and capitalist), an insignificant minority thrives by exploiting and oppressing the majority, i.e. the working people. Irreconcilable interests of the poor and the rich classes give rise to strife between them. The irreconcilable class struggle results in transition from one formation to another. The class struggle is the motive force of the development of antagonistic socio-economic formations, while social revolution as a culmination of class struggle is the law which determines the transition from a lower formation to a higher one.

Each formation features specific forms, content and character of labor (discussed in more detail below). At the same time, all the socio-economic formations were Influenced by forces whose effect is universal? The division and cooperation of labor ranks first among such forces.

As the division of labor becomes ever more pronounced, labor has an increasingly important role to play in the rise of man. In the long run, the division of labor in all its aspects serves to increase the labor productivity of man as society's main productive force. The division of labor promotes the accumulation of labor experience, improvement of skills, and greater competence and knowledge.

The division of labor as the isolation of various types of labor activity can take place on two levels – within the entire society and within an individual enterprise.

The intrasocietal division of labor manifests itself in the existence of various branches of production (industry, agriculture, transport, construction, etc.). For their part, these can be further divided and subdivided. For instance, industry is divided into mechanical engineering, ferrous metallurgy, chemical, food, textile, and many other industries.

The division of labor within individual enterprises implies the division of labor between people of various professions and trades – steel-markers, miners, carpenters, actors, etc.

Division of labor is an important aspect of the socialization of labor. What is meant by the socialization of labor is the economic ties conditioned by the social division of labor between people in the process of production, exchange, distribution, and consumption of the social product. As productive forces and productive relations develop, economic ties grow ever more complicated.

Specialization of production is a most important form of the social division of labor. It is manifested in the growth of the number of individual, independent branches of production which specialize in producing certain types of goods.

Specialization conduces to the mass-scale production of batches of goods, which, in turn, allows for the application of highly productive modern equipment and for bringing in new processes and more efficient forms of organizing labor and production. In a word, specialization is a most important prerequisite for scientific and technological progress.

The development of human society is accompanied by the growing division and cooperation of labor.

Cooperation of labor is a form of social labor involving a considerable number of people who interact with one another as they perform identical or interconnected operations.

Cooperation of labor has great advantages over individual, non-cooperated labor, in that it makes it possible to perform a large volume of work and complete construction projects within a short time.

Both the division and cooperation of labor are historical phenomena. As economic ties among the producers grow more and more complicated, there emerges the need for the cooperation of labor on the scale of the entire national economy.

Like cooperation of labor, division of labor has two aspects to it – the technico-economic and socio-economic. The former involves the cooperation of machinery, technology, labor, and material and financial resources. This aspect of cooperation is linked to the development of society's productive forces. The socio-economic aspect is conditioned by the form of property in the means of production. We shall first consider the technical and economic aspects, that is, the impact of machinery and technology on the content of labor.

The content of labor is conditioned by the means of labor and raw materials used as well as by the worker's operations and the product being manufactured. The development of the implements of labor is the main factor causing changes in the content of labor operations. In order to understand the way, the content of labor changes as the implements of labor are improved, suffice it to compare the work of a navy digging a ditch with a spade or a hoe and that of a suction-dredge operator, or the work of a miner wielding a pick and that of a cutter-loader operator.

The content of labor changes as different types of raw materials and stock are used. For instance, the use of agricultural machinery and aircraft, mineral fertilizers, and herbicides alleviates the work of the farmers and makes it increasingly similar to industrial labor.

The content of labor changes not only with the improvement of the means of labor, but also with the development of man's ability to perform increasingly complicated operations and ever more complicated tasks. The content of labor changes from generation to generation as labor becomes more diversified. Thus, as labor acquired increasing importance in the development of human society and in the division of labor, there emerged – apart from hunting and cattle-breeding – farming, spinning, weaving, mental-working, pottery, and navigation. Apart from trade and handicrafts, there appeared art and science. Mental work became distinct from physical work and the town became separated from the countryside. The content of labor changes not only with the accumulation of experience and improvement of skills, but also with the rise in the workers' educational and training levels.

The content of labor also depends on the scope of the use of new machinery and technologies, which reduce the amount of monotonous manual work.

From the point of view of its content, labor can be simple or complicated, mental or manual, creative or non-creative, managerial or subordinate.

Simple labor, for instance, the work done by shop hands or loaders, does not call for special training or high qualification.it is also referred to as unskilled labor.

By contrast, turners, fitters, adjusters, operators, etc., have to undergo more or less prolonged training before they start to work. Skilled labor requiring special training is referred to as complicated labor. It is in fact multiplied simple labor.

Under capitalism, this accounts for the further exacerbation of the contradiction between the town and the countryside. In rural areas, simple, less industrialized labor is subjected to increasingly intensive exploitation by the capitalist town with its factories and banks, skilled labor force, and higher labor productivity.

In a sense, the contraction between "the town," i.e. the developed major imperialist states, and "the countryside," i.e. the developing countries of Asia, Africa, and

Latin America (or, as it is often described today, the contradiction between the North and South) has developed into a grave economic crisis. The neocolonialist policy pursued by big monopolies, inequitable international economic relations, and the staggering foreign indebtedness estimated at 1 trillion US dollars have brought many of the developing states to disaster. The inequitable terms of trade with developed capitalist countries result in the price of the developing states' goods – minerals, sugar, meat, bananas, and other traditional export commodities – tumbling down and the prices of imported machinery and equipment soaring.

In 1984, the Latin American countries alone suffered a loss of 20 billion US dollars as a result of the deterioration of trade terms. In this context, indebtedness and the aggravation of the contradiction described above appear unavoidable.

As for mental and manual labor, their separation from each other and further independent development are a form of the most profound social division of labor. Man's labor organically combines the workings of the brain and of the muscles into a single process whereby man acts upon nature and transforms it. However, the combination and interaction of man's mental and physical activity in the process of labor is complicated and contradictory and is fraught with the possibility of their separation in the course of the productive forces' development.

At a certain stage in the development of the productive forces, mental activity, as a process of mastering the laws that govern the development of nature and society, isolates itself from the process of labor and starts evolving independently in the form of science, art, etc. Before the isolation takes place, mental activity remains a component of labor and its evolution is hindered by the predominance of physical labor that consumes all of man's time and strength. The less developed is labor and the lower is its productivity, the more physical effort is spent by society for biological reproduction (achieved by satisfying man's minimal requirements for food, clothing, and shelter), the less strength and time is left for any other activity, above all, mental activity.

The division of labor of a kind when the majority of people are predominantly engaged in physical labor and a minority (usually, these are members of the ruling classes) are engaged in mental labor (management, administration of state affairs, science and the arts) generates a contradiction between mental and physical labor, as individuals engaged in mental work exploit those engaged In physical work.

The development of the elements of mental activity in labor conduces to the latter's evolution into a creative process. In the process of creative labor, the worker spontaneously makes non-trivial decisions consciously seeking to fulfill his assignment. To be able to do so, the worker must be well trained and well versed in the latest achievements of science and technology. Non-creative labor is always monotonous, pre-programmed as it were. Assembly-line production is an example of non-creative labor.

Managerial labor is usually creative because the manager has to display initiative in organizing his subordinates' work and assigning them their jobs. By contrast, subordinates' labor is narrower and more specialized. It depends on other people's instructions and is therefore less creative.

As to the socio-economic aspect of the division of labor, it wholly depends on the form of ownership of the means of production and has decisive effect on the character of labor.

This deserves a more detailed discussion.

We have already mentioned that the implements of labor – tools, machinery, equipment, etc. – are the key element of productive forces. Their development level is the measure of man's mastery of nature and of the development of labor power, i.e. the productive ability of individual people and society as a whole. Moreover, the implements of labor constitute the most mobile element of productive forces. Each historical stage in the development of production is distinguished, above all, by the degree of development of the implements of labor.

The way people work – collectively or individually – depends on the instruments, i.e. whether they are intended for collective or for individual use, predetermines the character of labor.

Consider an example. The primitive-social character of man's earliest tools predetermined the need for collective labor under the primitive-communal system. The common use of tools, i.e. common labor, explains why the form of appropriation of the means of production was collective and why the first historical form of ownership of the means of production was communal, group ownership. This form made it impossible for some people to appropriate the means of production and for other people to be deprived of them. Thus, it ruled out the exploitation of man by man.

Man improves his tools in the process of production as he seeks to turn out as many products as possible and thereby satisfy his requirements to an ever-greater extent. The improvement of tools under the primitive-communal system led to the individualization of tools. Individual tools are the ones more suitable for individual use.

The individual, specialized character of the instruments of labor – the plow, the potter's wheel, the spindle, the bow, as well as the corresponding type of labor – individual labor – called for a change in the form of appropriation of means of production, for a transition from their collective appropriation to private appropriation, i.e. to private ownership of the means of production. This transition was inevitable because both production and man's manner of involvement in it had changed.

Thus, the character of labor depends on who – private owners or society as a whole – possesses the key means of production – land, factories, railways, airlines, etc. – on who controls their use in the process of labor. As relations of property in the means of production develop, the nature of relations among people in the process of labor changes, too.

Thus, the character of labor is closely linked to the development of the social system, each social system being characterized by a form of labor specific to it.

In the primitive-communal system, the tools (stones and sticks) man used to procure his scanty means of subsistence were extremely primitive. Group ownership and equalizing distribution made it possible to compensate for the production failures of some by the successes of others and thus to satisfy, on the whole, the minimum requirements of all. For instance, the food procured by the members of a commune either collectively or individually was placed at the disposal of the entire commune. The tools used in hunting and fishing were also disposed of by the entire tribe or clan. The Labor of each was labor for all.

The primitive-communal system persisted throughout a long period in the development of human society when the social division of labor was growing ever more pronounced. Farming, cattle breeding, crafts, and trade evolved into independent spheres of material production. The transition from fruit-gathering and hunting to the growing of crops, fruit, and vegetables and to the breeding, crafts, and trade evolved into independent spheres of material production. The transition from fruit-gathering and hunting to the growing of crops, fruits, and vegetables and to the breeding of domesticated animals marked a qualitative leap forward in human development. The transition from the use of primitive tools to the manufacture of more complicated ones was an essential factor in the development of labor at that stage of mankind's progress.

One should distinguish between live and objectified labor. Live labor is man's purposeful activity aimed at producing certain material goods. Objectified labor is man's past labor, which has been expended on turning out the means of production (machinery, equipment, buildings, fuel, raw materials, etc.).

The proportion of objectified labor in the process of production grows with the development of productive forces. Correspondingly, the owners of the means of production exert a growing influence on the relations among people in the process of production, exchange, distribution, and consumption. The owner of the means of production has a stake in acquiring more of them so as to secure for him a larger amount of material and cultural benefits. Therefore, he makes those who do not own any means of production work for him.

The emergence of private ownership of the means of production made the division of labor even more pronounced. An increasing amount of goods was produced for exchange for other goods rather than for personal consumption by the producer. The owners of the means of production specialized in making definite types of goods. That made the market essential to people's vital activity.

With the development of exchanges, labor acquired a two-fold character. The point is that the emergence of exchange called for the need to determine what quantity of goods was equivalent to a certain quantity of other goods. People learned to compare labor expended to produce various commodities regardless of their content.

Labor which is objectified in a commodity (provided that what is produced is a commodity or a service required by other people) is called abstract labor. Abstract labor lies in the basis of the cost of a commodity, that is, it determines the amount of money or other goods offered for the commodity at the market.

At The same time, any kind of labor is concrete as it yields use value, that Is, the properties of the commodity which are required by other people.

As labor acquired a two-fold character, private labor became distinct from social labor. In the final analysis, any production involves dozens of people of various trades and professions. Things are made because somebody needs them. This means that the labor of each commodity producer is part of social labor and is social by nature. However, in a society based on private ownership, commodity producers are disunited working on isolation from one another. Therefore, their labor, although essentially social, takes on the form of private labor. Its social character manifests itself through exchange.

It is on this basis that the contradiction between private and social labor emerges, and the product of man's labor comes to dominate him. Relations among commodities. This contradiction is only revealed on the market, when the problem of the sale of commodities arises. The product's grip on man is made even tighter because the commodity's value is determined by the socially necessary expenditure of labor rather than by the individual expenditure of labor. The socially necessary expenditure of labor is measured by the working time required to produce a commodity under the average social conditions of production, that is, at the average level of technology, producer's skill, and intensity of labor. As a rule, the socially necessary expenditure of labor depends on the production conditions under which the bulk of the given type of commodities is produced.

The emergency of private property and the development of commodity exchange led to the division of the primitive tribe into the poor and the rich. The impoverished individuals found themselves in debt slavery to the rich and were thus forced to work for the latter. By waging aggressive wars, turning the captives into slaves, and selling their fellow-tribesmen into slavery by way of exacting debts, the rich established their full sway over the enslaved poor. The primitive-communal socio-economic formation was thus replaced by the slave-holding system.

The slave-holding system was the first socio-economic formation in the history of mankind to be based on private property and the alienation of labor. This system was characterized by the slave-holders' private ownership of both the means of production and the workers, that is, slaves.

The alienation of labor originated when some groups of people started to appropriate the product of other groups' labor. The labor of the people who were deprived of the means of production and of the product of their labor became alienated from them. That accounts for their indifference and even aversion to labor. Slaves, being the property of individual or collective owners, possessed neither the means of production nor the means of subsistence.

Just as the results of their labor the slaves were the property of the owner. Their exploitation was achieved through outright violence, that is, by way of extra-economic coercion. The intrinsic contradiction of the slave-holding system stemmed from the coercive manner whereby the producers (slaves) came into contact with the means of production. As they were not interested in the result of their labor and in displaying a solicitous attitude to the implements of labor, the slaves resisted any attempts to make them work harder. They regarded the exploitation. Slave labor, in its entirety, took the form of unpaid work done for the slave-holder.

Under the slave-holding system, the content and character of labor were changed. Stone and wooden tools were replaced by implements made of copper, bronze, and, later, iron. However, the slave-holders preferred to issue their slaves with the most primitive implements of labor because the latter, hating their forced labor, took no care of expensive and complicated tools, misplacing, losing, or damaging them. The overwhelming majority of slaves served as a mere source of muscular energy in various branches of production (farming, cattle-breeding, construction, transportation, mining, etc.). They did arduous manual work. Only the ruling classes could engage in the administration of state affairs in science and art. Mental labor became opposed to physical labor.

In Antiquity, the upswings in economic and cultural development were the outcome of the cruel exploitation of slaves. The fruit of slave labor. It was with the arduous labor of huge numbers of slaves that canals and roads were laid out, large tracts of land irrigated, mines worked, large workshops established, and bridges, dams, pyramids, palaces, and temples constructed.

Various branches of human knowledge evolved. The slave-holding society produced such famous figures in culture and science as Mozi in China, Aristotle and Pythagoras in Greece, Kalidasa in India, and Cicero in Rome. For all that, the potential for the use of labor under the slave-holding system was limited as the slaves worked to the crack of overseers' whips.

Brimming with hatred for their owner and full of aversion for their forced labor, slaves would now and again rise against the exploiters. Slave labor gradually ceased to promote economic development. As for the slave-holders, they regarded physical labor as something unworthy of freemen, a situation when even the poor refused to work and expected sops ("bread and circuses") from the rich.

Idleness and labor are incompatible. There came a time in the development of the slave-holding system when the predominant production relations began to hinder the development of the productive forces. The irreconcilable contradiction between the productive forces and the production relations resulted in the downfall of the slave-holding system. The emergence of colony heralded the advance of another mode of production, feudalism. The colonies were former slaves or ruined peasants to whom land and implements of labor were leased. They had a stake in producing abundant crops because they retained a share of them. By special legislation, the colonists were bound to the land, deprived of the freedom of movement, the land, deprived of the freedom of movement, the right to serve in the army, etc. This system exhibited some of the features which were to become typical of the feudal socio-economic formation.

The slave-holding system gave way to feudalism. A higher development level of the productive forces under feudalism caused changes in the content of labor. There emerged new Sectors of the economy. In urban areas, the methods of processing raw materials were gradually improved, and crafts became more and more specialized. In 1600, the belt loom was invented in Europe. The 15th century saw the invention of the blast furnace, the compass, the mechanical watch, and book printing.

Improved methods of farming (the two- and three-field systems) were developed. New crops – potatoes, tomatoes, tobacco, and sunflower imported to Europe from America – became widespread. The instruments of labor were improved. The iron plow and other agricultural implements made of iron were used on a growing scale. In their drive for mastery over nature, people acquired vast experience and new skills.

However, feudal technology, based as it was on the large-scale application of manual labor, developed at a comparatively slow rate. The majority of the peasants, who were not free, used primitive tools.

Nonetheless, the character of labor was changing. The division and cooperation of labor become more pronounced. There emerged specialized branches of agriculture, such as butter manufacturing, bee-keeping, and gardening.

Under feudalism, commodity production acquired a new dimension and was concentrated in towns. With trade ties established among the hinterland regions of countries, national Markets began to take shape. The development of commodity production, based on the social division of labor and intended for sale, for the market, and for the exchange of products among various households, bolstered the role of labor in the improvement of the social form of labor.

Under feudalism, the social form of labor had a greater potential for increasing the role of labor in the rise of man as society's main productive force. In contrast with the slaves, the peasants and craftsmen were interested, to a degree, in the results of their labor. They owned some means of production and were thus able to engage in independent economic activity – to till land, breed cattle, or follow crafts – whenever they were free from quilt-rent and corvee, i.e. work on the land owned by the feudal lords, attending to the latter in their castles, and performing other compulsory service.

In terms of time, the serfs' necessary labor was distinct from their surplus labor, i.e. unpaid labor appropriated by the feudal lord by virtue of his being the owner of the land, water, woods, etc. The peasants, who were not freemen, were compelled to work for the feudal lord not so much because of their economic dependence on him as for the reasons of direct extra-economic coercion, which manifested itself in their being subject to their lord's will and jurisdiction. The feudal lord could at any time sell or give away his peasants with or without the land they were bound to. Feudal landownership, and the peasants' dependence on the landlords for the land to till, determined the character of the exploitation of labor under feudalism. Effected in many different ways, it was embodied in the feudal rent, which absorbed the whole of the peasants' surplus labor. Various duties and exactions left the peasants with the minimum means of subsistence.

The antagonisms inherent in feudal society continuously widened the gap between mental and physical labor, town, and country.

The peasants were unwilling to put up with their subjugated position. The epoch of feudalism abounded in peasant's uprising.

The crisis of the feudal mode of production became especially pronounced when the social form of labor based on feudal landownership and the peasant's being in personal bondage to the landlords fettered the development of productive forces. The production of commodities intended for exchange was growing in scale. As the market expanded, the more or less substantial commodity producers started to hire ruined farmers and craftsmen to work for them. This meant that in the womb of the feudal society there originated the capitalist mode of production based on capitalist private ownership and the exploitation of hired labor.

LABOR'S HIERARCHY OF NEEDS

The traditional concepts of systems analysis are applicable to project and work processes (Peacock & Badiru, 2023). The definitions of a project system and its components are presented next.

System. A project system consists of interrelated elements organized for the purpose of achieving a common goal. The elements are organized to work synergistically to generate a unified output that is greater than the sum of the individual outputs of the components.

Program. A program is a very large and prolonged undertaking. Such endeavors often span several years. Programs are usually associated with particular systems. For example, we may have a space exploration program within a national defense system.

Project. A project is a time-phased effort of much smaller scope and duration than a program. Programs are sometimes viewed as consisting of a set of projects. Government projects are often called *programs* because of their broad and comprehensive nature. Industry tends to use the term *project* because of the short-term and focused nature of most industrial efforts.

Task. A task is a functional element of a project. A project is composed of a sequence of tasks that all contribute to the overall project goal.

Activity. An activity can be defined as a single element of a project. Activities are generally smaller in scope than tasks. In a detailed analysis of a project, an activity may be viewed as the smallest, practically indivisible work element of the project. For example, we can regard a manufacturing plant as a system. A plant-wide endeavor to improve productivity can be viewed as a program. The installation of a flexible manufacturing system is a project within the productivity improvement program. The process of identifying and selecting equipment vendors is a task, and the actual process of placing an order with a preferred vendor is an activity.

The emergence of systems development has had an extensive effect on project management in recent years. A system can be defined as a collection (or partnership) of interrelated elements brought together to achieve a specified objective. In a management context, the purposes of a system are to develop and manage operational procedures and to facilitate an effective decision-making process. Some of the common characteristics of a system include:

1. Interaction with the environment
2. Objective
3. Self-regulation
4. Self-adjustment

Below is a formal general definition of a system:

> A collection (or partnership) of interrelated elements, whose collective output together (in unison) is higher than the mere sum of the individual outputs of the elements.

Representative components of a project system are the organizational subsystem, planning subsystem, scheduling subsystem, information management subsystem, control subsystem, and project delivery subsystem. The primary responsibilities of project analysts involve ensuring the proper flow of information throughout the project system. The classical approach to the decision process follows rigid lines of organizational charts. By contrast, the systems approach considers all the interactions necessary among the various elements of an organization in the decision process.

The various elements (or subsystems) of the organization act simultaneously in a separate but interrelated fashion to achieve a common goal. This synergism helps to expedite the decision process and to enhance the effectiveness of decisions. The supporting commitments from other subsystems of the organization serve to counterbalance the weaknesses of a given subsystem. Thus, the overall effectiveness of the system is greater than the sum of the individual results from the subsystems.

The increasing complexity of organizations and projects makes the systems approach essential in today's management environment. As the number of complex projects increases, there will be an increasing need for project management professionals who can function as systems integrators. Project management techniques can be applied to the various stages of implementing a system as shown in the following guidelines:

1. **Systems definition.** Define the system and associated problems using keywords that signify the importance of the problem to the overall organization. Locate experts in this area who are willing to contribute to the effort. Prepare and announce the development plan.
2. **Personnel assignment.** The project group and the respective tasks should be announced, a qualified project manager should be appointed, and a solid line of command should be established and enforced.
3. **Project initiation.** Arrange an organizational meeting during which a general approach to the problem should be discussed. Prepare a specific development plan and arrange for the installation of needed hardware and tools.
4. **System prototype.** Develop a prototype system, test it, and learn more about the problem from the test results.
5. **Full system development.** Expand the prototype to a full system, evaluate the user interface structure, and incorporate user training facilities and documentation.
6. **System verification.** Get experts and potential users involved, ensure that the system performs as designed, and debug the system as needed.
7. **System validation.** Ensure that the system yields expected outputs. Validate the system by evaluating performance level, such as percentage of success in so many trials, measuring the level of deviation from expected outputs, and measuring the effectiveness of the system output in solving the problem.
8. **System integration.** Implement the full system as planned, ensure the system can coexist with systems already in operation, and arrange for technology transfer to other projects.
9. **System maintenance.** Arrange for continuing maintenance of the system. Update solution procedures as new pieces of information become available. Retain responsibility for system performance or delegate to well-trained and authorized personnel.
10. **Documentation.** Prepare full documentation of the system, prepare a user's guide, and appoint a user consultant.

Systems integration permits sharing of resources. Physical equipment, concepts, information, and skills may be shared as resources. Systems integration is now a major concern of many organizations. Even some of the organizations that traditionally compete and typically shun cooperative efforts are beginning to appreciate the value of integrating their operations. For these reasons, systems integration has emerged as a major interest in business. Systems integration may involve the physical integration of technical components, objective integration of operations, conceptual integration of management processes, or a combination of any of these.

Systems integration involves the linking of components to form subsystems and the linking of subsystems to form composite systems within a single department and/or across departments. It facilitates the coordination of technical and managerial efforts to enhance organizational functions, reduce cost, save energy, improve productivity, and increase the utilization of resources. Systems integration emphasizes the identification and coordination of the interface requirements among the components in an integrated system. The components and subsystems operate synergistically to optimize the performance of the total system. Systems integration ensures that all performance goals are satisfied with a minimum expenditure of time and resources. Integration can be achieved in several forms including the following:

1. **Dual-use integration:** This involves the use of a single component by separate subsystems to reduce both the initial cost and the operating cost during the project life cycle.
2. **Dynamic resource integration:** This involves integrating the resource flows of two normally separate subsystems so that the resource flow from one to or through the other minimizes the total resource requirements in a project.
3. **Restructuring of functions:** This involves the restructuring of functions and reintegration of subsystems to optimize costs when a new subsystem is introduced into the project environment.

Systems integration is particularly important when introducing a new work into an existing system. It involves coordinating new operations to coexist with existing operations. It may require the adjustment of functions to permit the sharing of resources, development of new policies to accommodate product integration, or realignment of managerial responsibilities. It can affect both hardware and software components of an organization. Presented below are guidelines and important questions relevant for work systems integration.

- What are the unique characteristics of each component in the integrated system?
- How do the characteristics complement one another?
- What physical interfaces exist among the components?
- What data/information interfaces exist among the components?
- What ideological differences exist among the components?
- What are the data flow requirements for the components?
- Are there similar integrated systems operating elsewhere?
- What are the reporting requirements in the integrated system?
- Are there any hierarchical restrictions on the operations of the components of the integrated system?
- What internal and external factors are expected to influence the integrated system?

- How can the performance of the integrated system be measured?
- What benefit/cost documentations are required for the integrated system?
- What is the cost of designing and implementing the integrated system?
- What are the relative priorities assigned to each component of the integrated system?
- What are the strengths of the integrated system?
- What are the weaknesses of the integrated system?
- What resources are needed to keep the integrated system operating satisfactorily?
- Which section of the organization will have primary responsibility for the operation of the integrated system?
- What are the quality specifications and requirements for the integrated systems?

Maslow's Hierarchy of Needs (Maslow, 1943) is very much applicable in any work environment. According to Maslow's theory, the five different orders of human needs are:

1. **Basic physiological needs:** Includes food, water, shelter, and the like. In modern society, the basic drives of human existence cause individuals to become involved in organizational life. People become participants in the organization that employs them. Thus, at the simplest level of human needs, people are motivated to join organizations, remain in them, and contribute to their objectives.
2. **Security and safety:** Security means many things to different people in different circumstances. For some, it means earning a higher income to assure freedom from what might happen in case of sickness or during old age. Thus many people are motivated to work harder to seek success, which is measured in terms of income. It can also be interpreted as job security. To some people, such as civil servants and teachers, the assurance of life tenure and a guaranteed pension may be strong motivators in their participation in employing organizations.
3. **Social affiliation:** An employee with a reasonable well-paying and secure job will begin to feel that belonging and approval are important motivators in his/her organizational behavior.
4. **Esteem:** The need to be recognized, to be respected, and to have prestige (self-image and the view that one holds of oneself). There is a dynamic interplay between one's own sense of satisfaction and self-confidence on one hand, and feedback from others in such diverse forms as being asked for advice on the other hand.
5. **Self-actualization:** The desire to become more and more what one is, to become everything that one is capable of becoming. The self-actualized person is strongly inner-directed, seeks self-growth, and is highly motivated by loyalty to cherished values, ethics, and beliefs. Not everyone reaches the self-actualized state. It is estimated that these higher-level needs are met about ten percent of the time.

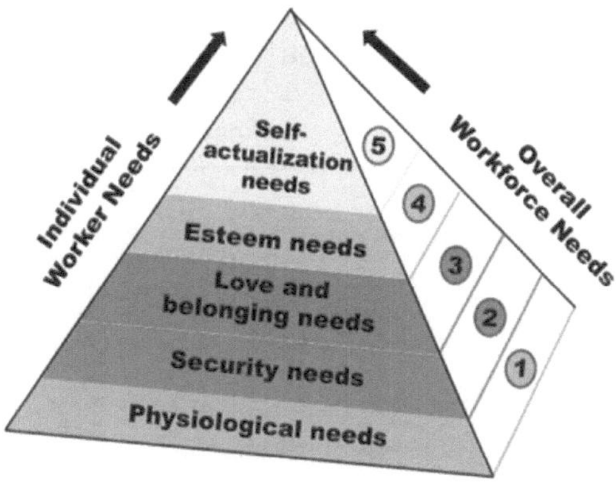

FIGURE 1.1 Multidimensional pyramid of needs of workers and the organization.

In any organization, the prevailing hierarchy of needs of the worker must be evaluated in the context of the organization's own hierarchy of needs. In this regard, Badiru (2012) developed an adaptation of the conventional triangle of the hierarchy into a multidimensional pyramid of needs work workforce development, as illustrated in Figure 1.1.

Soft Skills for Economic Development

Not only is work essential for personal and organizational advancement, but it is also essential, from a synergistic systems perspective, for national social and economic development. The gross domestic product (GDP) is the eventual coalescing of work done at various levels of the nation. GDP is the monetary value of all the finished goods and services produced within a country in a defined period of time. Though GDP is usually calculated on an annual basis, it can be calculated on a quarterly basis as well in order to increase the granularity of management policies and practices to increase the national output. GDP includes all private and public consumption, government spending, investments, and exports (minus imports) that occur within a national boundary. In other words, GDP is a broad and composite measurement of a nation's overall economic activity. It can be calculated as follows:

$$\text{GDP} = \text{C} + \text{G} + \text{I} + \text{NX}$$

where

C = all private consumption, or consumer spending
G = the sum of government spending
I = sum of all the country's investment, including corporate capital expenditures
NX = the nation's total net exports, calculated as total exports minus total imports

GDP is commonly used as an indicator of the economic health of a country, as well as a gauge of a country's standard of living. If a country's standard of living is high, the workers in the country do well for themselves. Since the mode of measuring GDP is uniform from country to country, GDP can be used to compare the productivity of various countries with a high degree of accuracy. Adjusting for inflation from year to year allows for an objective comparison of current GDP measurement trends. Thus, a country's GDP from any period can be measured as a percentage relative to previous years or quarters. Consequently, GDP trends can be used in measuring a nation's economic growth or decline, as well as in determining if an economy is in recession, which has a direct impact on workers and the work environment.

THE COLOR OF LABOR

With respect to the hierarchy of needs of people, it is interesting to note the hierarchy of labor in the workplace with respect to the color of helmets. Figure 1.2 illustrates the different colors of helmets of "labor" in the workplace. It represents a sort of who is who in the hierarchy of functions and needs in the work environment. This helps

Helmet Color	Helmet Shape	Function Description
Yellow		Labourer, Heavy-duty operations, & Construction tasks
Grey		Site visitors
Red		Firefighters
Brown		Welders, high heat operations
Blue		Electricians and Technical operators
Green		Safety officers
Pink		female workers* *in some companies it is used as an additional helmet
White		Managers, Engineers, Supervisors, Foreman

FIGURE 1.2 The color of labor helmet in the workplace.

us to close this chapter and move on to more strategic underpinnings of the theme of this book.

REFERENCES

Badiru, A. B. (1993), **Managing Industrial Development Projects: A Project Management Approach**, Van Nostrand/John Wiley, New York, NY.

Badiru, A. B. (2019), **Project Management: Systems, Principles, and Applications**, 2nd Edition, Taylor & Francis CRC Press, Boca Raton, FL.

Badiru, A. B., and O. A. Omitaomu (2007), **Computational Economic Analysis for Engineering and Industry**, Taylor & Francis CRC Press, Boca Raton, FL.

Badiru, A. B., O. Ibidapo-Obe, and B. J. Ayeni (2019), **Manufacturing and Enterprise: An Integrated Systems Approach**, Taylor & Francis/CRC Press, Boca Raton, FL.

Clark, C., N. Hewitt, J. Brown, and D. Jaffe (2008), **Who Built America? Working People and the Nation's History**, 3rd Edition, Bedford/St. Martin's, New York, NY.

Maslow, A. H. (1943), "Theory of Human Motivation," ***Psychological Review***, Vol. 50, No. 4, pp. 370–396.

Peacock, B., and A. Badiru (2023), **Industrial Engineering in Systems Design: Guidelines, Practical Examples, Tools, and Techniques**, Taylor & Francis CRC Press, Boca Raton, FL.

Savchenko, P. (1990), What Is Labour?, **Special Report of the Institute of Economy, Russian Academy of Sciences**, Progress Publishers, Moscow, Russia.

2 Systems View of Global Manufacturing

INTRODUCTION

The entire world is a global system of systems. We are all interconnected through one medium or another. Global manufacturing is, indeed, one of the most influential systems in the world. Given the way we are intertwined and interconnected around the world, we must understand our interdependency. This book presents practical systems-based strategies that nations, cities, and communities can adopt and/or adapt for developing and sustaining manufacturing for the purpose of sustainable industrialization. The contents are broadly pertinent for all nations, developed or developing, as a path toward economic development through manufacturing.

The economic development and industrialization of any nation depend on solid foundations of manufacturing. Over the passage of time, industries come and go and requirements for economic commerce shift. Nations that survive and continue to thrive use adaptive strategies that meet current and future needs. If good sustainable strategies are available, organizations and communities can survive any turbulence that develops in the normal course of economic development. This book is operationally focused, with adaptive contents to fit the needs of different global audiences. This book uses the case and template of global trade relations to demonstrate how modern technology innovation strategies can be developed and harmonized to keep a nation moving forward economically. Manufacturing technology transfer models and techniques are applicable to other technology transfer relationships between any pair of collaborators. Trade and commerce between any two entities are similar to any country-to-country relationship. For example, within the Africa-USA trade relationship reside country-by-country interfaces of the USA with specific countries of Africa. Some examples include USA-Nigeria, USA-Liberia, USA-South Africa, USA-Sudan, USA-Ghana, and USA-Egypt bi-lateral relationships. The sub-Saharan countries of Africa are most noticeable when the issues of developing and underdeveloped countries are covered in the literature or popular press. The economic size of the USA puts it on a comparative par with the continent of Africa. There are several global issues affecting how nations interact politically, socially, militarily, economically, and commercially. Bhargava (2006) addresses many of the global issues faced by citizens around the world.

CASE EXAMPLE OF TECHNICAL WORKFORCE DEVELOPMENT

The case example in this section is based on the Africa-USA International Conference on Manufacturing Technology (MANUTECH) organized between 1993 and 2004 by a group of Nigerian engineers in Diaspora (Nnaji et al, 1993, 1994, 1996, 1997,

DOI: 10.1201/9781003537755-2

2000; Badiru et al, 2002, 2004; Okogbaa et al, 2007). The conference convened on a rotational basis between locations in Africa and the USA. MANUTECH 2007 exists only in a planning document (Okogbaa et al, 2007) as the last conference was planned, but did not convene due to wide professional reassignments of the organizers at that time. The topics addressed by the MANUTECH series are still germane today for the purpose of technology transfer strategies between countries of the world.

At the conference, engineers, scientists, managers, researchers, chief executives, administrators, and students from across Africa and the USA participated in bilateral technology-related discussions. The economic advancement being experienced by Africa in the current global economic climate is exactly what the MANUTECH conference organizers envisioned in January 1993. It has taken two decades, but slow and steady wins the race. Africa's 2014 economic rise is a testimony that solid intellectual foundations bear fruit in the long run. The MANUTECH 2004 welcome statement below aptly conveys this optimism. This could serve as a template for other inter-country engagements.

Today, the concept of MANUTECH continues to represent an international congregation of scholars and practitioners interested in the advancement of Manufacturing Technology all over the world. Forward and reverse manufacturing technology transfers between the USA and other countries are imperatives for global competitiveness in the new world marketplace for these partners. The collaborative theme epitomizes a triangular relationship between academia, industry, and government establishments. This facilitates pragmatic approaches to transforming classroom theories into practical applications for the betterment of industry. Through the conference of the 1990s and the early 2000s, the knowledge-driven pursuits of academia were coupled with the profit-driven objectives of industry to provide a framework for the policy-driven strategies of government agencies. The wide variety of topics befitting the theme of the conference and the ensuing intellectual pursuits paved the way for several subsequent constructive pursuits by the participants and formed the impetus for the theme of this present book.

The problems of developing countries, such as those in Africa, cannot be successfully addressed until those countries are able to implement appropriate technologies for infrastructure development and for basic manufacturing. In particular, the adaptations of existing and appropriate manufacturing technologies are essential to sustainable industrial growth and development.

Economic, social, and political progress in Africa can be enhanced by the creation of local industries which provide employment for the growing population, cater to the needs of the population for local consumer goods, provide export products, and stimulate foreign investment. This book provides a forum for the exchange of information and ideas between professionals in developing countries and those in the Western world for manufacturing technology transfer strategies. Such exchanges could form the basis for creating better awareness of the role of manufacturing technologies in the context of the development and economic stability of underdeveloped countries and could also become the catalyst for increased and sustained dialogue on those issues.

The Appropriate Technology

Ibidapo-Obe (1996) addresses the question of whether we can develop models that will measure the effectiveness or appropriateness of technologies, keeping in mind all the facets and modes of local and national aspirations. He contends that we are in a period of rapid social and economic change, much of which is driven by science and technology. Around the world, waves of science and technology-based innovation are leading to a multitude of new products, processes, and services. Science and technology, in effect, are merging as the basis for national comparative advantage. Whether a country produces technology or uses it strategically is dependent on an environment that is rich in specialized infrastructure, specialized technology, financial resources, educational programs, management practices, marketing services, and highly skilled workers. To be a part of the change process, every community must develop a "technology" engine to link with this rapid and evolutionary development. Further, Ibidapo-Obe asks, "who will be the players in the community technology engine?" He provides suggestions that include the following:

- The technology sector.
- Every local industry and business, including general business associations.
- Educational, research, and scientific institutions.
- Planners and officials from all levels of government.
- Individuals involved in technology transfer, such as brokers, institutional representatives, and consulting engineers.
- Financiers, bankers, and local investors.
- Labor groups.
- Elected officials, political representatives, and government officials.

The entities identified above are responsible for providing the leadership and guidance that communities need to fire up the technology engine. We need to embrace information technology, which has become the engine of our modern society. Information technology is the pathway for achieving development objectives. We can now attempt to quantify and, thereafter, input all the above variables into the process of developing the technology input-output model. As a strategy, Ibidapo-Obe (1996) proposes a procedure to understand the workings of the various human sensors and activators (neuro-control mechanisms), especially vision and the brain. These will serve as a micro-model for the larger world systems, provided we can establish a one-to-one relationship between the micro (neuro-control) model and the macro (world development) system. The basic idea is to introduce some formalities in the restructuring of vagueness in many decision problems. Some of the considerations in this development process include the following:

1. Inputs: Knowledge, Capital, Materials, Hardware, Software, and Human Resources.
2. System: Engineering, Production, Personnel, Manufacturing, Management, and Facilities.
3. Output: Better products and services. Higher profits through technology applications.

Engineering is the connecting arm of technology in the development between the current scenario and the desired future state. The products of the modern world only function as long as large parts of society operate according to a laid-out plan. Unfortunately, in many developing countries, things often don't go according to plan. Underutilized equipment, rusting machinery, and idle factories, working at less than half of their designed capacities, provide cogent testimonies of the dilemma in developing countries. In all the above, having a solid global situational awareness is essential.

Everything is industrially interconnected in the modern marketplace. It is a systems world, and all organizations must demonstrate a systems view of the world when managing international projects. Teamwork and mutual understanding take on a different flavor when different cultures are involved. Not only must we think globally but we must also act globally through culturally sensitive project management. It is impossible for any organization to run any large project nowadays without some aspect of international involvement.

USING THE VEE SYSTEMS MODEL

Figure 2.1 illustrates the Vee Systems Model applied to manufacturing. It shows that several factors revolve around the manufacturing lifecycle. A decomposition of the overall manufacturing system facilitates collaborative manufacturing involving different players from different parts of the world. A good example is the globally

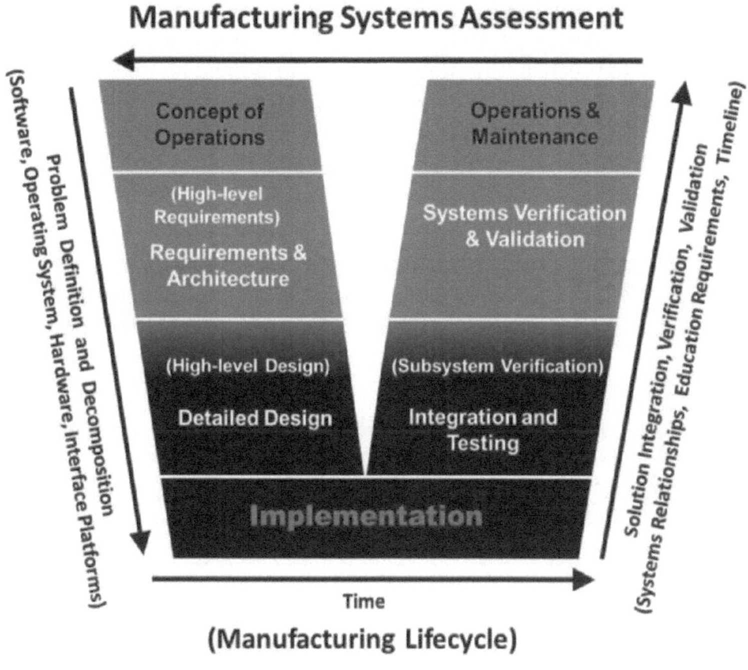

FIGURE 2.1 Vee systems engineering model

connected network of manufacturers from around the world involved in building Boeing's 780 Dreamliner airplane. This interconnectivity means that manufacturers must work with industrial collaborators spanning different cultures, languages, religions, customs, trade practices, skills, and politics. The conglomeration of manufacturing works for Boeing because the company uses a global systems engineering approach. It is through the systems approach of industrial engineering that manufacturing enterprises could become more efficient. In this context, the quote below aptly describes the systems impact in a military operational context:

"Aerospace engineers build weapons, civil engineers build targets, and industrial engineers make sure that the weapons get to the right targets." – Adedeji Badiru, AFIT

MANUFACTURING PROJECT MANAGEMENT

Manufacturing project management is the pursuit of organizational goals within the constraints of time, cost, and performance expectations. The principles and techniques of conventional project management are applicable to the management of technology transfer efforts. Project expectations can be in terms of physical products, services, and results. Global interactions of business and industry imply that global situational awareness be instituted for international project management. Situational awareness is the process of recognizing and appreciating unique factors that define the operating characteristics in a given geographical location. These factors can range widely from one country to another. What are believed to be normal operating conditions in home country operations may be taboo, illegal, restricted, or forbidden in another country. Far too often, organizations don't pay enough attention to this fact. Training programs, briefings, seminars, and sensitivity role plays are often used to prepare personnel for overseas assignments. But the fact remains that global awareness failure points still exist among organizations operating overseas, away from home base – whether from East to West or from West to East or elsewhere.

In world news, headlines abound on topics of global project interconnections in oil, autos, industrial credits, construction, and so on. Such mega deals are not limited to the G8, G10, G12, or G20. Even the little non-G economies are in the mix of international projects. What connects all of these is global project management.

HYBRIDIZATION OF CULTURES

The increased interface of cultures through international project outsourcing is gradually leading to the emergence of hybrid cultures in many developing countries. A hybrid culture derives its influences from diverse factors, where there are differences in how the local population views education, professional loyalty, social alliances, leisure pursuits, and information management. A hybrid culture poses a big challenge to managing internationally outsourced projects. The problem can be mitigated by a strategic program of situational awareness. Table 2.1 presents some of the pros and cons of cultural enmeshing at the global level.

TABLE 2.1
Pros and Cons of Global Hybrid Cultures

Pros	Cons
Global awareness for competition.	Subjugation of one culture to the other.
Culturally diverse workforce.	Cultural differences in work ethics.
Access of international labor force.	Loss of cultural identity.
Intercultural peace and harmony.	Bias and suspicion.
Facilitation of social interaction.	Confusion about cultural boundaries.
Closing of trade gaps.	Non-uniform business vocabulary.

MARKET COMPETITION

Many western industries cannot compete globally because of labor costs, where the pursuit of competitiveness is often directed. The competitive advantage for many manufacturers will come from appropriate infusion of technology into the enterprise. Strategic research, development, and implementation of technological innovations will give manufacturers the edge needed to successfully compete globally. Despite many decades of lamenting about the future of manufacturing, very little has been accomplished in terms of global competitiveness. Part of the problem is the absence of a unified project management approach. Managing global and distributed production teams requires a fundamental project systems approach.

One recommendation is to pursue more integrative linkages of technical issues of production and the operational platforms available in industry. Many concepts have been advanced on how to bridge the existing gaps. But what is missing appears to be a pragmatic project-oriented roadmap that will create a unified goal that adequately, mutually, and concurrently addresses the profit-oriented focus of practitioners in industry and the knowledge-oriented pursuits of researchers in academia. These problems embody both scientific and management issues. Many researchers have not spent adequate time in industry to fully appreciate the operational constraints of industry. Hence, there is often a disconnection between what research dictates and what industry practice requires. An essential need is the development of a global project roadmap. One aspect that is frequently ignored in this respect is the set of human-cultural factors operating in globally distributed work teams. A culturally sensitive project management approach will enable an appreciation of this crucial component of global projects.

GLOBAL AWARENESS REQUIREMENT

To enhance global awareness, the preparation approach that can be more effective and sustainable is to address specific sets of questions without making any prior assumptions in each and every overseas assignment case. Even previous international experience can become out of phase because local situations can change over time. The questions categorized below can be helpful.

Sample questions for Pre-assignment Planning

- What is the organizational mission in the overseas location?
- Are the personnel aware of the global situation as it relates to the country or region of operation?
- Will the assignment be as a team, individual operation, or organizational attachment?
- What legal status will be associated with the project during the overseas assignment?
- What country locations are involved in the assignment?
- What healthcare situations and medical services are available during the assignment?
- What personal family arrangements and precautions are relevant for the overseas assignment?
- What prudent precautions are possible? Will update? Health insurance? Medical records?
- What is the currency exchange rate? Location and access to bureau de change?

Sample questions for local conditions

- What kind of local environment will be in effect as the assignment location? Urban city? Rural? Jungle? Mountain? Desert?
- What is the typical weather pattern? Seasons? Cold? Hot? When, where, and how long?
- What are the general living conditions? Access to clean water? Bathroom facilities?
- What are the domestic transportation options? Cost? Reliability? Access? Rivers? Trails? Trains? Shuttle buses? Commuter planes?
- What is the level of environment consciousness?
- What is the local social hierarchy?

Sample questions for cultural nuances

- Is there a compilation of acceptable and unacceptable behavior in the assignment location?
- What are the political beliefs?
- What are the typical customs?
- What are the prevailing religious beliefs and restrictions?
- Are there guidelines for use of body language and gestures?
- What are the offensive symbols and words?

Sample questions for geographic location

- What are the bordering regions, areas, and/or countries to the assignment location?
- What is the operating terrain? Flat? Hilly? Road system?
- Is there access to rivers, oceans, beaches, etc.?
- What is the vegetation like?
- What are the major industries? Manufacturing factories? Service industry?
- Are there farming establishments? Access to fresh foods?

Sample general questions

- What is the social environment?
- Is there night life? Is it accessible? Is it safe?
- What are the crime statistics?
- What mode of law enforcement?
- What is the restriction on currency importation and exportation?
- What is the local manpower level? Skills? Availability? Cost? Dependability?
- What is the educational opportunity? Access? Affordability? Level?
- What international support organizations are available? Red Cross?
- What is the human rights record of the locality?

If questions such as the above are addressed forthrightly, international projects can run more effectively, more efficiently, and more successfully. Typical sources of barriers to a comprehensive global awareness include the following:

Denial: This presents the danger of not fully recognizing the problem of global awareness. A person may dismiss the problem or minimize the gravity of the problem.

Past Experience: This has the danger of luring the personnel into an erroneous decision based on a past or similar experience.

Complacency: This has the danger of feeling of over-confidence about knowing what obtains in the assignment location, thereby dismissing the necessity for proactive preparation.

Insufficient information: This has the danger of misinterpretation due to limited or unreliable information about the assignment location.

OVERSEAS LABOR COSTS

Some cultures, by their inherent nature, offer lower labor costs for international operations. Consequently, the search for a competitive operating site might take an organization to a culture that is totally different from the base station. The pervasiveness of fast routing of information makes it easy for a business to find a seemingly receptive overseas site to relocate operations. The consequence of such relocation is a transfer of culture in either direction, often with limited situational awareness. Many organizations don't fully appreciate the differences in the operating cultures. They pay attention only to the physical and economic aspects of their operations. More often than not, the cultural shock and unsuccessful assimilation from either end of the culture transfer can lead to project failures.

Many of the economically underserved countries in Asia, Africa, Oceania, and Latin America are frequent targets of international project development. Those countries have common characteristics, such as highly dependent economies (devoted to producing primary products for the developed world), traditional, rural social structures, high population growth, and widespread poverty. Certain characteristics exist that may constitute barriers to successful global projects. Some of these are:

- Limited access to information (substandard telecommunications infrastructure).
- Politically induced trade barriers.

- Cultural norms impede the free flow of information.
- Existing and abundant supply of cheap, albeit untrained, workforce.
- Orientation of manpower toward artisan and apprenticeship labor.

GLOBAL WORKFORCE CONSTRAINTS

Most project outsourcing points are in developing and underdeveloped nations. These locations often have repressive cultures that are replete with norms that the Western world would find unacceptable. A cultural bridge usually is missing between the developed nations and the developing nations with respect to workforce capabilities. Thus, there are increasing cultural and economic disparities between global business partners. Some of the local issues to be factored into global situational awareness programs include:

- Poverty.
- Pollution.
- Disease.
- Inferior health services.
- Political oppression.
- Gender biases.
- Economic and financial scams.
- Wealth Inequities.
- Social permissiveness among the elite.

Project personnel posted to these regions are shocked by the level of cultural differences that they experience. In some cases, they maintain a "laissez faire" and hands-off attitude. But there have also been cases where some of the personnel take advantage of the loose culturally acceptable social contacts that could be to the detriment of an international project.

HIERARCHY OF NEEDS IN A DIFFERENT CULTURE

The psychology theory of "Hierarchy of Needs" proposed by Abraham Maslow in his 1943 paper, "A Theory of Human Motivation," still governs how different cultures respond to the dimensions of global project expectations. A culturally induced disparity in the hierarchy of needs implies that we may not be able to fulfill our project responsibilities along the global spectrum of international projects. In a culturally different workforce, the specific levels and structure of the needs may be drastically different from the typical mode observed in the Western culture. Maslow's hierarchy of needs consists of five stages:

Physiological Needs: These are the needs for the basic necessities of life such as food, water, housing, and clothing (i.e., Survival Needs). This is the level where access to money is most critical.

Safety Needs: These are the needs for security, stability, and freedom from physical harm (i.e., Desire for a safe environment).

Social Needs: These are the needs for social approval, friends, love, affection, and association (i.e., Desire to belong). For example, social belonging may bring about a better economic outlook that may enable each individual to be in a better position to meet his or her social needs.

Esteem Needs: These are the needs for accomplishment, respect, recognition, attention, and appreciation (i.e., Desire to be known).

Self-Actualization Needs: These are the needs for self-fulfillment and self-improvement (i.e., Desire to arrive). This represents the stage of opportunity to grow professionally and be able to selflessly help others.

In an economically underserved culture, most workers will be at the basic level of physiological needs, and there may be cultural constraints on moving from one level to the next higher level. This fact has an implication on how cultural interfaces may fail between host and guest nations involved in a global project.

PROJECT SUSTAINABILITY

Sustainability isn't just for the environment. Project sustainability on the global front is as much of a need as the traditional components of project management, spanning planning, organizing, scheduling, and control. For international projects, the preemption of cultural conflicts is far better than post-occurrence remedies. Proactive pursuit of project best practices can pave the way for project success on a global scale.

Cultural infeasibility is one of the major impediments to project outsourcing in an emerging economy. The business climate of today is very volatile. This volatility, coupled with cultural limitations, creates problematic operational elements for global projects located in a developing country. For example, an inquiry or revelation of personal information is viewed as taboo in many developing countries. Consequently, this may impede the collection, storage, and distribution of workforce information that may be vital to the success of global project management. Global situational awareness and its best practices are essential for any organization engaged in international projects. In these days of globalization, nothing happens without international cooperation, and cooperation cannot succeed without effective situational awareness on each side of the cooperating partners.

REFERENCES

Badiru, A. B., B. O. Nnaji, G. Okogbaa, and P. Egbelu (July 8–11, 2002), "MANUTECH Conference Opening Address," **Proceedings of the Sixth Africa-USA International Conference on Manufacturing Technology**, MANUTECH 2002, Abuja, Nigeria.

Badiru, A. B., B. O. Nnaji, G. Okogbaa, and P. Egbelu (July 12–14, 2004), "Welcome Address for Africa-USA Manufacturing Collaboration Conference," **Proceedings of the Seventh Africa-USA International Conference on Manufacturing Technology**, MANUTECH 2004, Port Harcourt, Nigeria.

Bhargava, V., editor (2006), **Global Issues for Global Citizens: An Introduction to Key Development Challenges**, The World Bank, Washington, DC.

Ibidapo-Obe, O. (August 12–15, 1996), "Models for Technology Appropriate for Sustainable Development," in Nnaji et al, editors, **Proceedings of the Third Africa-USA International Conference on Manufacturing Technology**, MANUTECH 1996, Accra, Ghana, pp. 193–194.

Nnaji, B. O., A. B. Badiru, G. Okogbaa, and P. Egbelu (January 11–14, 1993), "Conference Overview," **Proceedings of the First Africa-USA International Conference on Manufacturing Technology**, MANUTECH 1993, Ikeja, Lagos, Nigeria.

Nnaji, B. O., A. B. Badiru, G. Okogbaa, and P. Egbelu (August 8–11, 1994), "Conference Overview," **Proceedings of the Second Africa-USA International Conference on Manufacturing Technology**, MANUTECH 1994, Ikeja, Lagos, Nigeria.

Nnaji, B. O., A. B. Badiru, G. Okogbaa, and P. Egbelu (August 12–15, 1996), "Conference Overview," **Proceedings of the Third Africa-USA International Conference on Manufacturing Technology**, MANUTECH 1996, Accra, Ghana.

Nnaji, B. O., A. B. Badiru, G. Okogbaa, and P. Egbelu (August 11–14, 1997), "Conference Overview," **Proceedings of the Fourth Africa-USA International Conference on Manufacturing Technology**, MANUTECH 1997, Pittsburgh, Pennsylvania.

Nnaji, B. O., A. B. Badiru, G. Okogbaa, and P. Egbelu (July 10–14, 2000), "Conference Overview," **Proceedings of the Fifth Africa-USA International Conference on Manufacturing Technology**, MANUTECH 2000, Abuja, Nigeria.

Okogbaa, G., A. B. Badiru, B. O. Nnaji, and P. Egbelu (July 18–20, 2007), "Conference Overview," **Planning Document for the Eighth Africa-USA International Conference on Manufacturing Technology**, MANUTECH 2007, Knoxville, Tennessee.

3 Manufacturing and Economic Development

INTRODUCTION

Manufacturing is a key foundation for national development. Industrialization is one of the primary means of improving the standard of living in a nation. Manufacturing provides the foundation for sustainable industrialization. Manufacturing refers to the activities and processes geared toward the production of consumer products. Industrialization refers to the broad collection of activities, services, resources, and infrastructure needed to support business and industry. History indicates the profound effect that the Industrial Revolution had on world development. Industrialization will continue to play an active role in national economic strategies. Nations that continue to seek international aid for one thing or another can lay a solid foundation for industrialization through manufacturing enterprises. A nation that cannot institute and sustain manufacturing will be politically delinquent and economically retarded in the long run. A good industrial foundation can positively drive the political and economic processes in any nation.

To achieve and sustain manufacturing development, both the technical and managerial aspects must come into play. This book contains the managerial processes necessary to facilitate manufacturing development. Project management and technology management are presented as viable means of achieving manufacturing development. Technology transfer is a key component of technology management.

COMPLEXITY OF MANUFACTURING

The economic and social problems facing many nations in the present technological age are very challenging. The gloomy trends have been characterized by the stagnant state of established industries, decline in productivity, closure of poorly managed corporations, globalization of markets, increased dependency on physical technology, increased apathy towards social issues, proliferation of organized and sophisticated illegal financial deals, requirement for expensive capital investments, and neglect of economic diversification endeavors. All these problems led to the decline of manufacturing in the 2000s. Fortunately, a resurgence of manufacturing is presently being experienced in different parts of the world.

An industrial development project is a complex undertaking that crosses several fields of endeavor. The diverse political, social, cultural, technical, organizational, and economic issues that intermingle in industrial development compound the manufacturing efforts. Sophisticated managerial approaches are needed to control the interaction effects of these issues. Financial power is necessary but not a sufficient requirement for industrial development. A common mistake by a developing economy is to simply dump scarce financial resources at a development problem without

DOI: 10.1201/9781003537755-3

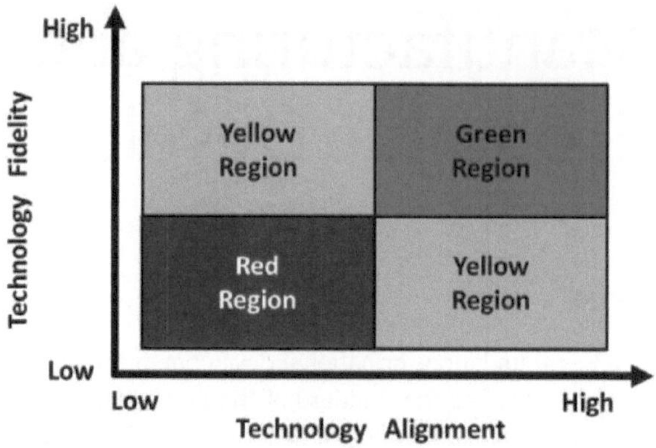

FIGURE 3.1 Technology transfer fidelity and alignment.

making adequate improvement in the management processes needed to support the development goal. Focusing on the technical aspects of a development project to the detriment of the managerial aspects will only create the potential for failure. A project management approach can facilitate an integrated understanding of the complex issues involved in industrial development and, thus, pave the way for success.

Contrary to general belief, industrial development problems are not limited to developing nations alone. Even in developed and industrialized nations, large pockets of industrially neglected communities can be found. Residents of these communities live in abject poverty, despite the ambient affluence of the overall nation. The countries that are most blessed with natural resources are often those that suffer the most from industrial neglect. The problem is typically that they do not know how to initiate and implement the projects that are needed to exploit the available resources. Some rely on misaligned technology transfers. As portrayed in Figure 3.1, technology aligned is essential for achieving sustainable development goals. If technology fidelity is high, but alignment is low, then the effectiveness of the transfer will be low (Red Region). If technology fidelity is high and alignment is high, then we have the greatest potential for success (Green Region).

Proper applications of project management techniques can facilitate further responsiveness and effectiveness to pave the way for the advancement of manufacturing.

STRATEGIC PLANNING FOR MANUFACTURING

Industrial planning determines the nature of actions and responsibilities required to achieve industrial development goals. Strategic planning involves the long-range aspects of manufacturing development efforts. Planning forms the basis for all actions. Strategic planning for manufacturing development can be addressed at three distinct levels as discussed below.

SUPRA-LEVEL PLANNING

Planning at this level deals with the big picture of how manufacturing development fits the overall and long-range needs of the community, region, or nation. Questions faced at this level may concern the potential contributions of manufacturing activities to the standard of living in the community, the development of resources needed to provide basic amenities in the community, the required interfaces between development projects within and outside the community, the government support for manufacturing development, responsiveness of the local culture, and political stability.

Macro-Level Planning

Planning at this level may address the overall planning within a defined industrial boundary. The scope of the development effort and its operational interfaces should be addressed at the macro-level planning. Questions addressed at this level may include industry identification, product definition, project scope, availability of technical manpower, availability of supporting resources, workforce availability, import/export procedures, development policies, effects on residential neighborhoods, project funding and finances, and project coordination strategies.

Micro-Level Planning

This level of planning deals with detailed operational plans at the task levels of manufacturing activities. Definite and explicit tactics for accomplishing specific development objectives should be developed at the micro level. Factors to be considered at micro-level planning may include scheduled time, training requirement, tools required, task procedures, reporting requirements, and quality requirements. Manufacturing is capital intensive. If not planned properly, the investment may prove to be an uneconomic venture.

INDUSTRIAL DEVELOPMENT AND ECONOMIC DEVELOPMENT

Industrialization is one side of the economic coin. Industrial development can directly translate to economic development if proper management practices are followed. Industrial development can be formulated as a basic foundation for economic vitality and national productivity improvement. Some of the major factors that positively impact interactions of industrial and economic processes include the following:

- Unification of national priorities.
- Diversification of the economic strategy.
- Strong strategy for development of rural areas.
- Adequate investment in research and development.
- Stable infrastructure to support development effort.
- Political stability that prevents economic disruption.
- Social and cultural standards that improve productivity.

Industrial development plays different types of roles in economic development. The structure of certain nations is such that the economic and industrial systems are on different development tracks. To facilitate the interaction of industrial and economic endeavors, plans must be made to allow the industrial system to coexist symbiotically with other production forces (Atta, 1981; Badiru, 2012; ASCE, 2013; Badiru, 2016; Badiru et al, 2019).

PURSUIT OF TECHNOLOGICAL CHANGE

Technological change is needed to drive industrial development. Technological progress calls for the use of technology. Technology, in the form of information, equipment, and knowledge, can be used productively and effectively to lower production costs, improve service, improve product quality, and generate higher output levels. The information and knowledge involved in this technological progress include those which improve the performance of management, labor, and raw materials.

Technological progress plays a vital role in improving total productivity. Statistics on developed countries, such as the United States, show that between 1870 and 1957, 90 percent of the rise in real output per man-hour can be attributed to technical progress. It has been shown that industrial or economic growth is dependent on improvements in technical capabilities as well as on increases in the amount of the conventional factors of capital and labor. Technological change is not necessarily defined by a move toward the most modern capital equipment. Rather, technological changes should be designed to occur through improvements in the efficiency of the use of existing equipment. The challenge to developing countries, such as those in Africa, is how to develop an infrastructure that promotes and utilizes available technological resources.

WORKFORCE SOCIAL CHANGE

Industrial development requires change. Society must be prepared for change in order to take advantage of new industrial manufacturing opportunities. Efforts that support industrial development must be instituted in every aspect of society and everything it does. If a society is better prepared for change, then positive changes can be achieved. Industrial development requires an increasingly larger domestic market. The social systems that make up such markets must be carefully coordinated. The socio-economic impact of industrialization cannot be overlooked. Social changes are necessary to support industrial development efforts. Social discipline and dedication must be instilled in society to make industrial changes possible. The roles of the members of society in terms of being responsible consumers and producers of industrial products must be outlined. People must be convinced of the importance of the contribution of each individual whether that individual is acting as a consumer or as a producer. Industrial consumers have become so picky that they no longer simply accept whatever is offered in the marketplace. In cases where social dictum directs consumers to behave in ways not conducive to industrial development, changes must be instituted. If necessary, acquired taste must be developed to like and accept the products of local industry.

To facilitate consumer acceptance, the quality of domestic industrial products must be improved to competitive standards. New technologies that facilitate increased industrial productivity must be embraced whether through domestic sources or through technology transfer. For an industrial product to satisfy the sophisticated taste of the modern consumer, it must exhibit a high level of quality and responsiveness to the needs of the consumer. Only high-quality industrial products and services can survive the increasing competitive market. Some of the approaches for preparing a society for industrial development changes are listed below.

- Highlight the benefits of industrial development.
- Keep citizens informed of the impending changes.
- Get citizen groups involved in the decision process.
- Promote industrial change as a transition to a better society.
- Allay the fears about potential loss of jobs due to industrial automation.
- Emphasize the job opportunities to be created by industrial development.

Stipulation for the use of local raw materials is one policy that may necessitate a change in social attitudes. Society (consumers and producers) must understand the importance of local raw materials in domestic manufacturing activities. When backed against the wall of production, producers must develop the necessary indigenous technology or processes for using local materials. The technology transfer modes presented in this book provide a guideline for a local adaptation of inwardly transferred technology.

THE ROLE OF EDUCATION SECTOR

A more aware population is a more responsive and supportive population. It is essential to include education, training, and public awareness in the strategic planning for manufacturing development. For example, inflation can adversely affect investments in manufacturing development. Inflation is a disease that feeds on itself, but not without the help of people. The basic attitude of citizens towards consumption and wealth greatly determines the trend of inflation. Everyone wants to buy and sell consumer products without paying attention to the production efforts needed to generate those products. The seller wants to sell at whatever exorbitant prices he or she can get. The naive buyer does not help the situation either. He or she is willing to buy whatever he or she wants at whatever the asking price even if it means sacrificing personal economic sense, which, in the long run, with many like-minded consumers, can jeopardize the economic path of the whole nation.

The producer, encouraged by high demand, shrewdly cuts production and, thus, reduces market supply, which then elevates consumer prices even further. The fact that a product is in short supply makes the consumer yearn for more of it. The product quickly becomes a status symbol. The few who claim to be the "class" of society and they are willing to pay whatever price to maintain that status quo. Each consumptive individual is unaware of the economic agony that reckless personal acts can cause for the larger population. It is not that consumers are unpatriotic. Rather, it is that they don't know the adverse implications of collective personal economic habits. This is why educating the public is an important aspect of combating inflation and paving the way for manufacturing development. The producer, the retailer, and the buyer all need to know the real causes of inflation, its boomerang effects, and its potential remedies. The government should play a very active role in the education process. The indoctrination process needs to start right at the grassroots. Children should be exposed to the concepts of battling inflation so that they can grow up to become responsible producers, retailers, or consumers. The early instillation of these concepts to the youths will even have the beneficial side effect of imbibed national pride, which will serve the social and political interests of the nation in later years. Most of the leadership problems in many countries can be traced to a lack of deep-rooted pride and interest in the national welfare.

In an underdeveloped nation, nothing is more discouraging than to hear people proclaim how they have given up on their nation. They claim that nothing works, no improvements are forthcoming, and that the government is unresponsive to the suffering of the people. How do we expect citizens who feel negative about their nation's performance when they, themselves, reach the position of national leadership? Certainly, attitudes and perception will not change overnight. If citizens carry negative impressions into higher offices, all they will do is to perpetrate their views by saying "didn't I say nothing works in this country?" Obviously, such people have not had early mechanisms of instilling national pride and economic dedication in them. This is why early education and public awareness of economic processes are essential for creating sustainable manufacturing activities.

Government programs cannot succeed without people's cooperation in the implementation process. Herein lies the irony of governing and being governed. We blame the government for not instituting corrective programs and we turn right around to take actions that innocently sabotage the prevailing programs, thereby compounding the burden of the government. People must be educated on the implications of their actions. Presenting a national strategy is not enough. Ensuring that citizens are educated about how to support and execute the strategy is equally essential. When a consumer product is in short supply, everyone complains and blames the system. No one reflects on the individual contribution at each consumer or producer level. Consider the case of a factory worker who buys a locally made product and complains of its inferior quality compared to imported brands. He or she forgets the personal unique position to contribute significantly to the quality of local products through on-the-job activities. There are many cases auto industry workers buying foreign-made vehicles not because of the need for diversity of assets, but rather due to the belief of the better quality of imported vehicles. This is like sending a signal that says, "I build the local vehicle, and I know about it. Please don't buy it!" That is a fallacy of production dedication. The worker bites the hand that provides the feeding.

To further illustrate the problem of lost dedication, consider this hypothetical scenario of an auto industry employee. As the demand for imported vehicles increases so do the prices of the vehicles. So, the employee asks the employer for higher wages so that a higher-end imported vehicle can be purchased. To pay those higher prices, the employer raises the prices of the locally made vehicles to make more money to pay the higher wages. When the prices of the local vehicles go up without a comparable increase in quality (thanks to the employee's work delinquency), the public buys even less of the locally made vehicles. So, the employer sells fewer cars and makes less money. In the end, the employee is laid off. This shows that the employee's economic education is lacking in the first place. This is why it is essential to broaden the sense of the economic awareness of consumers.

The scenarios painted here demonstrate the fact that the economic system needed to support manufacturing development is very complex. Individual actions, as innocent and minor as they may appear, can go a long way in advancing or impeding the overall economy. We should be convinced that the education of the public is a key factor in solving many economic problems before industrial development can take root.

POLITICAL LANDSCAPE

Politics determines the crux of economic vitality and economic vitality determines the crux of politics. With the changing political environments in developing nations, as those in Africa, it is difficult to stabilize government policies. Many governments are in perpetually transient states. Just when policies are beginning to take a firm hold, a new administration comes in and, as a show of new power, overturns all previous achievements. Billions of dollars have been wasted by succeeding governments through the absurd practice of abandoning projects started by preceding administrations regardless of merit. For the sake of national progress, governments should make a commitment to retain and execute worthwhile projects irrespective of who started them. Foresight should also be exercised when embarking on new industrial projects to ensure that subsequent governments can find the merit for their continuation. Laying a solid foundation for an industrial development project can facilitate a lasting coexistence of political, economic, and industrial activities. The major role of government in industrial development must be that of a facilitator. This is particularly important in inter-government negotiations for mutual development. The interdependence of the political, economic, and industrial systems is depicted in Figure 3.2. It is noted that manufacturing vitality advances along the manufacturing system axis depending on the advances in the economic and political systems. The building blocks of industrial performance are represented by the industrial performance surface and are dependent on the positive interplay of manufacturing, economic, and political systems.

If the economic and political processes are at low levels, no significant industrial development can be expected. As the economy picks up and the political atmosphere

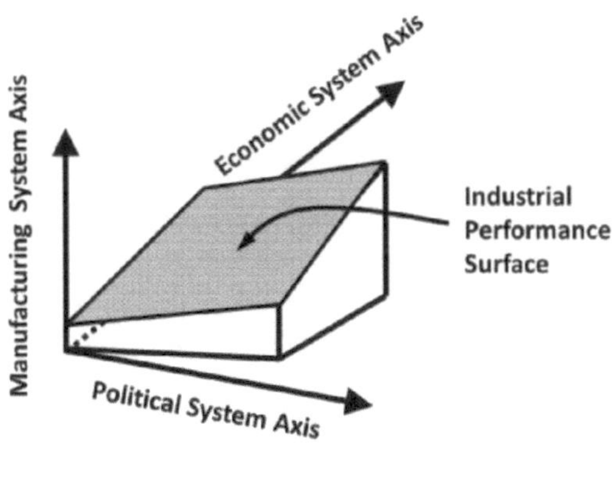

FIGURE 3.2 Interdependence of political, economic, and industrial systems.

improves, the level of industrial development will begin to increase. To maximize industrial output, political and economic systems must operate at symbiotically high levels. This is why a developing nation must recognize that chaotic political situations will adversely affect overall national development efforts.

GLOBAL INFLUENCES

The new world economic order calls for new approaches to development. The prevailing changes around the world will have profound effects on international trade and domestic manufacturing activities. The changes in eastern Europe, the advancements in western Europe, the merging of nations, the breakup of nations, and the emergence of Africa as a viable market will all affect international trade in the coming years. No nation can be insulated from what goes on in other nations. If not impacted directly, impact can occur through intermediary trading partners. The mobility of manufacturing outputs and products will be one common basis for global trade communication. Companies and countries must recognize the trend and refocus efforts and direct investments accordingly. Some of the key aspects of globalization that impact industrial development activities include the following:

- Transition of some countries being trade partners to being trade competitors.
- Reduction of production cycle time to keep up with the multi-lateral introduction of new products around the world.
- Increased efforts to reduce product life cycle from years to months.
- Increased responsiveness to the needs of a mixed workforce.
- The pressure to mitigate the adverse impacts of cultural barriers.
- Relaxation and expansion of trade boundaries.
- Tightening of trade relationships.
- Politically driven trade sanctions.
- Integration of operations, services, and consolidation of efforts.
- More effective and responsive communication, including social media.
- Increased pressure for multi-national cooperation.
- Need for multi-company and multi-product coordination.
- National security barriers.

The interdependence of world projects in modern times makes it imperative that global development efforts be pursued. Political, industrial, economic, and social disasters in one nation can easily spill over to other nations and create adverse chain reactions. Refugee problems now plaguing many nations are cruel reminders that nations can no longer exist in isolation.

There are several unique factors that impinge on industrialization. Some of these factors are addressed in the following sections. While technology continues to push the limit of perfection in developed nations, citizens of underdeveloped nations continue to be relegated to antiquated production tools. Industrial development has unique aspects that must be considered in formulating development strategies. Some of these aspects are discussed below.

RURAL AREA DEVELOPMENT

Because of their underdeveloped state, rural areas offer significant potential for industrial development. Ironically, it is that same underdevelopment that makes them susceptible to neglect. Productive members of rural areas move to urban areas in large numbers in search of a better life. The more technological advancement in a nation, the more rural residents migrate to urban areas. Consequently, the manpower needed to support the development of a rural area is often not consistently available. The key to keeping the manpower in their localities is the development of the localities and the creation of employment opportunities and the provision of the basic amenities of life. Rural neglect has created a highly uneven distribution of population in most countries. Urbanization without industrialization is doomed to economic failure.

The sparse population in rural areas means large land areas with potential for industrial setups are available at a relatively low cost. This makes them particularly attractive for the location of new plants. However, investors often shy away from the rural areas because of the limited availability of supporting infrastructure such as electrical power, water, communication, raw materials, and transportation facilities. If power, water, transportation, and communication are guaranteed, the development of the rural areas will be affected successfully. In addition to providing these services, a nation should offer incentives (e.g., tax incentives) to rural investors. Some key elements of rural area development include the following.

- Improvement of basic infrastructure.
- Creation of investment incentives for local businesses.
- Establishment of academic institutions geared to local educational needs.
- Establishment of technology extension services through academic institutions.
- Creation of a local clearing house for technology information.
- Government assistance for local product development.
- Provision of adequate rural health care services.
- Cultural Barriers to Industrialization.

Cultural barriers can inhibit industrial development. Most industrialization programs are fueled by foreign ideas, technologies, and approaches. In some cultures, cloaks of protection have been developed to protect society from foreign influences. The greatest fear often involves the potential moral decadence that may accompany industrial development as people migrate from one nation to another in response to the development process. Cultural barriers may also prohibit certain segments of society from participating fully in the development process. These cultural barriers must be removed to pave the way for full participative industrial development. Some cultures have evolved to a level of indifference to fraudulent and corrupt practices that are detrimental to development efforts. In some cases, corruption is viewed as an acceptable way of doing business. Culturally permissive attitudes to corruption must be altered before general development can take hold. Loyalty to the overall development goals rather than individual pursuits is required as a part of cultural changes for industrial development.

EDUCATION FOR WORKFORCE DEVELOPMENT

He who owns the knowledge controls the power. Both formal and informal education should play a vital role in national industrial development. In this modern day of high technology, adequate education is necessary to succeed in any work environment. Even in some industrialized nations, a large percentage of the adult population in neglected communities is functionally illiterate. It used to be that the children of these poverty-stricken communities were necessary to drive the wheels of manual labor in local factories. But the present and future industries, with an increasing push for automation, will not need many of the services of the labor-intensive workforce. Education geared towards the new industrial direction is necessary to participate actively in industrial development. Poor parents always proclaim that they don't want their children to end up where they did. Yet, no drastic educational efforts have been made to ensure that they don't.

RESEARCH AND DEVELOPMENT PARTNERSHIP

Research and education go hand-in-hand in creating pathways for industrial development. Partnerships between government, university, and industry offer good avenues for addressing the pressing issues of national development. Figure 3.3 illustrates a multifaceted approach to government-university-industry partnerships. There is enough

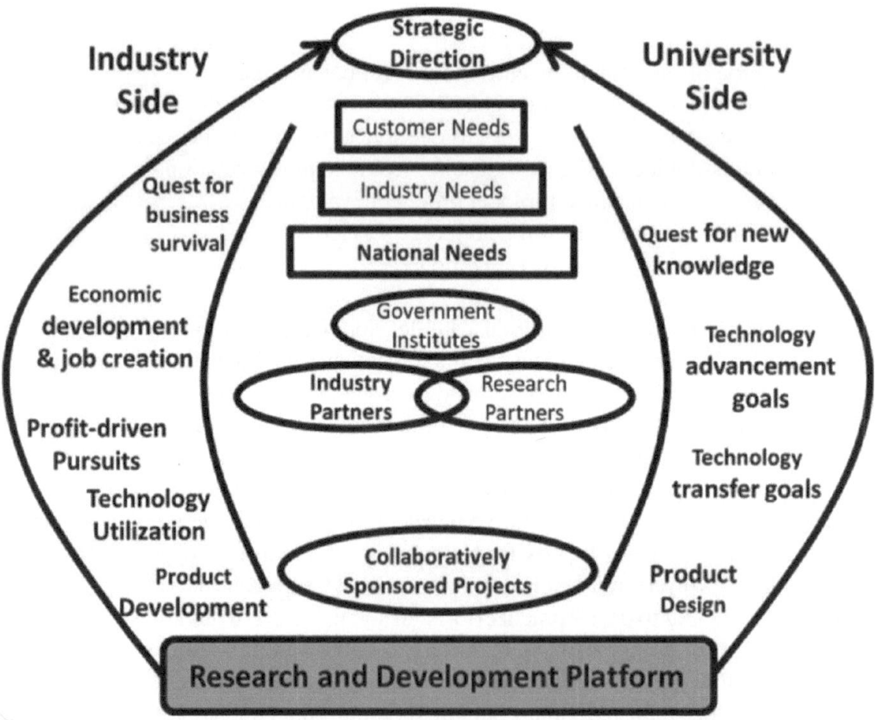

FIGURE 3.3 Multi-tiered partnership of government, industry, and university.

room for each entity's goals. On the side of business and industry, the end goal is to maximize profit and ensure survival. On the side of the university, the primary goal is the quest for new knowledge, technological advancement, and technology transfer. On the part of the government, the goal is to increase the wealth and security of the nation. Specific research-based partnerships can address multi-lateral goals. Quantitative tools such as multi-attribute decision modeling are of interest for this purpose.

Knowledge is a sustainable capital. The establishment of a formal process for the interface of institutions of higher learning and industry can be one of the capitals for industrial development. Universities have unique capabilities that can be aligned with industry capabilities to produce symbiotic working relationships. Private industrial research projects must complement public industrial research programs. Academic institutions have a unique capability to generate, learn, and transfer technology to industry. The quest for knowledge in academia can fuel the search for innovative solutions to specific industrial problems. A collaborative industry is a fertile ground for developing prototypes of new academic ideas. Industrial settings are good avenues for the practical implementation of technology. Industry-based implementation of university-developed technology can serve as the impetus for further efforts to develop new technology. Technologies that are developed within the academic community, mainly for research purposes, often languish in the laboratory because of the lack of formal and coordinated mechanisms for practical industrial implementation. The potentials of these technologies go untapped for several reasons, including the following.

- The researcher does not know which industry may need the technology.
- Industry is not aware of the technology available in academic institutions.
- There is no coordinated mechanism for technical interface between industry and university groups.

Universities interested in technology development are often hampered by the lack of adequate resources for research and training activities. Industry can help in this regard by providing direct support for industrial groups to address specific industrial problems. Universities also need real problems to work on as projects or case studies. Industry can provide these under a cooperative arrangement. The respective needs and capabilities of universities and industrial establishments can be integrated symbiotically to provide benefits for each group. University courses offered at convenient times for industry employees can create opportunities for a university-industry interaction. Class projects assigned to industry employees can be designed to address real-life industrial problems. This will help industry employees to create focused and rewarding projects. Class projects developed in the academic environment can be successfully implemented in actual work environments to provide tangible benefits. With a mutually cooperative interaction, new developments in industry can be brought to the attention of academia while new academic research developments can be tested in industrial settings. The growing proliferation of Distance Learning (DL) and Distance Education (DE) programs can be leveraged to spread educational opportunities to other industries.

The role of the government should be the facilitator to provide linkages and programs that enable multi-organizational collaboration. One desirable approach is for the government to establish and fund technology clearing houses. Academic institutions can serve as convenient locations for technology clearing houses. Such clearing houses can be organized to provide up-to-date information for industrial development activities. Specific industrial problems can be studied at the clearing house. The clearing house can serve as a repository for information on various technology tools. Industry would participate in the clearing house through the donation of equipment, funds, and personnel time. The services provided by a clearing house could include a combination of the following.

- Provide consulting services on technology to industry.
- Conduct on-site short courses with practical projects for industry.
- Serve as a technology library for general information.
- Facilitate technology transfer by helping industry identify which technology is appropriate for which problems.
- Provide technology management guidelines that will enable industry to successfully implement new technology in existing operations.
- Expand training opportunities for engineering students and working engineers.

The establishment of centers of excellence for pursuing industrial development-related research is another approach to creating a favorable atmosphere for industry-university interaction. As an example, from 2000 to 2006, co-author Adedeji Badiru founded and directed the Industrial Engineering Center for Industrial Development Research (IE-CIDER) within the Department of Industrial Engineering at the University of Tennessee in Knoxville, Tennessee, USA. The center was dedicated to research studies involving the multi-dimensionality of factors and processes affecting the movement of a product from concept stage to the commercialization stage with specific ties to regional economic development. A part of the services of the center was industry location feasibility studies.

AGRICULTURE AND INDUSTRIALIZATION

A hungry society cannot be an industrially productive society. It is generally believed that an underdeveloped economy is characterized by an agricultural base. Based on this erroneous belief, several developing nations have abandoned their previously solid agricultural base in favor of alternate means of industrialization. The fact is that a strong agricultural base is needed to complement other industrialization efforts. Agriculture is a viable source of industrialization when we consider the broad definition of industrialization presented earlier in this book. Mechanized farming is a sort of industrialization. If industrialization does not yield immediate benefits, society will be exposed to the double jeopardy of hunger and material deprivation. Once abandoned, agriculture is a difficult process to regain. Since agricultural processes take several decades to perfect, the revitalization of abandoned agriculture may require several decades. Agriculture should play a major role in the foundation of industrial development. The agricultural sector can serve as a viable market for a concomitant industry through supply chain interfaces. It is interesting to note that the agricultural revolution of the past paves the way for the subsequent industrial revolution.

Human history indicates that humans started out as nomad hunters and gatherers, drifting to wherever food could be found. About 12,000 years ago, humans learned to domesticate both plants and animals. This agricultural breakthrough allowed humans to become settlers, thereby spending less time wandering in search of food. More time was, thus, available for pursuing stable and innovative activities, which led to the discoveries of better ways of planting and raising animals for food. That initial agricultural discovery eventually paved the path to the agricultural revolution. During the agricultural revolution, mechanical devices, techniques, and storage mechanisms were developed to aid the process of agriculture. These inventions made it possible for more food to be produced by fewer people. The abundance of food meant that more members of the community could spend time with other pursuits rather than the customary labor-intensive agriculture. Naturally, these other pursuits involved the development and improvement of the tools of agriculture. The extra free time brought on by more efficient agriculture was used to bring about more technological improvements in agricultural implements. These advanced agricultural tools led to even more efficient agriculture. The transformation from the digging stick to the metal hoe is a good example of the raw technological innovation of that time. With each technological advance, less time was required for agriculture, thereby allowing more time for further technological advancements. The advancements in agriculture slowly led to more stable settlement patterns. These patterns led to the emergence of towns and cities. With central settlements away from farmlands, there developed a need for transforming agricultural technology to domicile technology that would support the new city life. The transformed technology was later turned to other productive uses which eventually led to the emergence of the industrial revolution. To this day, the entwined relationships between agriculture and industry can still be seen. Figure 3.4 shows a representation of the intersection of manufacturing technologies.

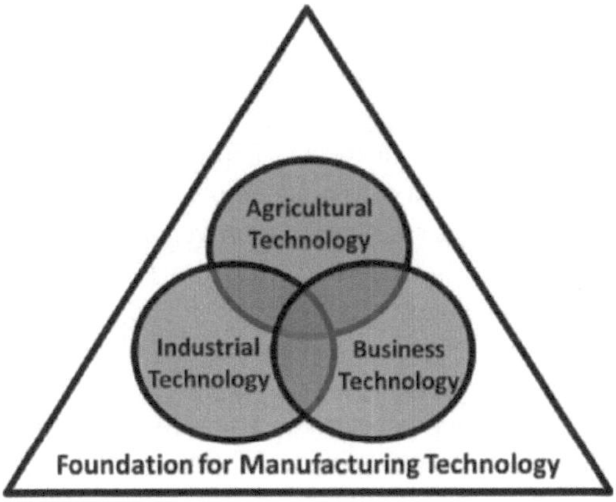

FIGURE 3.4 Intersection of manufacturing technologies.

WORKFORCE FOR INDUSTRIALIZATION

Technical, administrative, and service workforce are necessary to support industrialization initiatives. People make development possible. No matter how technically capable a machine may be, people will still be required to operate or maintain it. Soon after World War II, it was generally believed that physical capital formation was a sufficient basis for development. That view was probably justified at that time because of the role machinery played during the war. It was not obvious then that machines without a trained and skillful workforce did not constitute a solid basis for development. It has now been realized that human capital is as crucial to development as physical capital. The investment in workforce development must be given a high priority in the development plan. Some of the important aspects of manpower supply analysis for industrial development include the following.

- Level of skills required.
- Mobility of the manpower.
- The nature and type of skills required.
- Retention strategies to reduce brain drain.
- Potential for coexistence of people and technology.
- Continuing education to facilitate adaptability to technological changes.
- Who is available to educate and train the workforce.

With respect to who may be available to contribute to the education and training of the national workforce, we offer the following perspectives. Military service can and has an impact on advancing workforce education. By virtue of having gone through structured education and training for their service requirements, ex-servicemen and ex-servicewomen have unique skills and experience that can be channeled toward educating and training the national workforce for specific needs. Military education can have far more impact on national advancement than we realize. In the days of Adedeji Badiru's own university studies, he marveled at the knowledge base, span of expertise, and professionalism of many of his engineering instructors. He wondered how such a high concentration of marvelous militarily experienced engineers could be found in one institution. He later found out that this was not an isolated incident in one institution. It turned out that many engineering professors in the 1960s and 1970s across the USA had served in the US Navy or Army during World War II. After the war, through government programs, many transferred their military training, education, and expertise into lecturing at universities. The positive impression that he had of his engineering professors gave him the early incentives to apply more forthright efforts to his engineering education and, subsequently, choose academia as his own career path. The consequence is that the foundational knowledge acquired from the military engineers-turned professors continues to serve students years later. The conclusion is that the military directly and indirectly influenced the advancement of technical workforce in the United States. What the US is experiencing today in terms of being a world leader is predicated on a foundation of consistent technical education over the years. Other nations, particularly developing nations, can learn from this example.

There is often a debate whether the military should continue to invest (and how much) in advanced military education. The fact is that the national investment in advanced military education is not only essential to keep the military on the cutting edge of warfare technology, but also to positively impact the national landscape of education that may contribute to the advancement of manufacturing technology. Recognizing the urgent need to address global societal issues from a technical standpoint, in 2008, the National Academy of Engineering (NAE, 2008) of the United States published the "14 Grand Challenges for Engineering." The challenges have global implications for everyone, not just the engineering professions. As such, solution strategies must embrace all disciplines. The military, by virtue of its global presence and wider span of involvement in advanced technical education, can provide the technical foundation for addressing many of these challenges. Engineers of the future will need diverse skills to tackle the multitude of issues and factors involved in adequately and successfully addressing the challenges. Military engineers are necessary to provide the diverse array of technical expertise, discipline, and professionalism required. Technical education, covering Science, Technology, Engineering, and Mathematics (STEM) at advanced levels, provides a sustainable opportunity for the military to impact the 14 grand challenges listed below.

1. Make solar energy economical.
2. Provide energy from fusion.
3. Develop carbon sequestration methods.
4. Manage the nitrogen cycle.
5. Provide access to clean water.
6. Restore and improve urban infrastructure.
7. Advance health informatics.
8. Engineer better medicines.
9. Reverse-engineer the brain.
10. Prevent nuclear terror.
11. Secure cyberspace.
12. Enhance virtual reality.
13. Advance personalized learning.
14. Engineering the tools of scientific discovery.

All the above have implementation implications for manufacturing technology, either directly or indirectly through transitional interactions. An extract from the NAE document on the 14 grand challenges reads: *"In sum, governmental and institutional, political and economic, and personal and social barriers will repeatedly arise to impede the pursuit of solutions to problems. As they have throughout history, engineers will have to integrate their methods and solutions with the goals and desires of all society's members."* Who is better capable to analyze, synthesize, and integrate than the technically trained military? Advanced military education is necessary and should be sustained to carry out this charge. A recommendation is that qualified ex-service people should be sent back to school for advanced technical education and turn them into future engineering educators so that future engineering students can benefit, and the entire nation can also benefit. This may even have the side benefit of having society view the military in the positive light of national economic advancement beyond national security.

TECHNOLOGY ATTRIBUTES

Some specific attributes of technology can facilitate industrial development, but technology must be managed properly to play an effective role. There is a multitude of new technologies that have emerged in recent years. Hard and soft technologies such as computing tools, cellular manufacturing, intelligent software tools, and social media are changing the landscape of manufacturing rapidly. But much more remains to be done in actual implementation. It is important to consider the peculiar characteristics of new technology before establishing adoption and implementation strategies for applications in industrial development. The justification for the adoption of a new technology should be a combination of several factors rather than a single characteristic of the technology. The important characteristics to consider include productivity improvement, improved product quality, reduction in production cost, flexibility, reliability, and safety.

An integrated evaluation must be performed to ensure that the proposed technology is justified economically and technically. The scope and goals of the proposed technology must be established right from the beginning of an industrialization project. This entails the comparison of industry objectives with the overall national goals in the areas discussed below.

Market Target: This should identify the customers of the proposed technology. It should also address items such as market cost of the proposed product, assessment of the competition, and the market share.

Growth Potential: This should address short-range expectations, long-range expectations, future competitiveness, future capability, and prevailing size and strength of the competition.

Impact on National Goals: Any prospective technology must be evaluated in terms of the direct and indirect benefits to be generated by the technology. These may include product price versus value, increase in international trade, improved standard of living, cleaner environment, safer workplace, and improved productivity.

Profitability: An analysis of how the technology will contribute to profitability should consider past performance of the technology, incremental benefits of the new technology versus conventional technology, and value added by the new technology.

Capital Investment: Comprehensive economic analysis should play a significant role in the technology assessment process. This may cover an evaluation of fixed and sunk costs, cost of obsolescence, maintenance requirements, recurring costs, installation cost, space requirement cost, capital substitution potentials, return on investment, tax implications, cost of capital, and other concurrent projects.

Skill and Resource Requirements: The utilization of resources (manpower and equipment) in the pre-technology and post-technology phases of industrialization should be assessed. This may be based on material input-output flows, high value of equipment versus productivity improvement, required inputs for the technology, expected output of the technology, and utilization of technical and nontechnical personnel.

Risk Exposure: Uncertainty is a reality in technology adoption efforts. Uncertainty will need to be assessed for the initial investment, return on investment, payback period, public reactions, environmental impact, and volatility of the technology.

National Productivity Improvement: An analysis of how the technology might contribute to national productivity may be verified by studying industrial throughput, efficiency of production processes, utilization of raw materials, equipment maintenance, absenteeism, learning rate, and design-to-production cycle.

INFRASTRUCTURE DEVELOPMENT

Industrialization built upon a solid foundation can hardly fail. The major ingredients for durable industrial, economic, and technological developments are electrical power, water, transportation, and communication facilities. These items should take priority in major industrial development projects. Of course, housing infrastructure for the workforce is another basic need that must be accounted for in an industrial development strategy. A four-sided manufacturing infrastructure model to provide a foundation for industrial development is presented in Figure 3.5. Such a development pursuit seeks to ensure the following amenities.

Primary Amenities:

- Reliable Power Supply.
- Consistent Water Supply.
- Good Transportation System.
- Efficient Communication System.

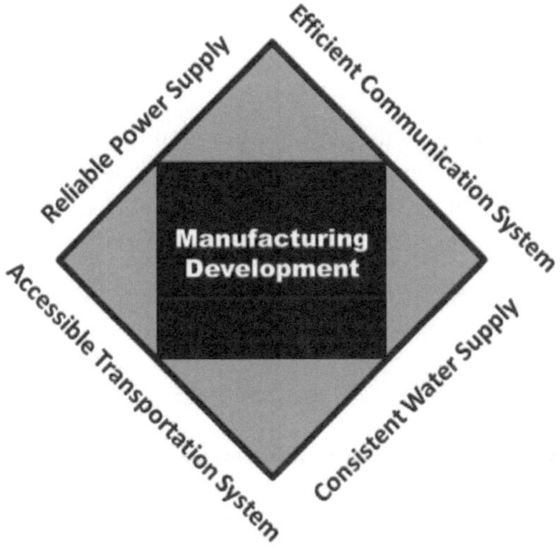

FIGURE 3.5 Four-Sided manufacturing infrastructure model.

Supporting Amenities:

- Housing.
- Education.
- Health Care.

The provision of adequate healthcare facilities is essential to building a strong industrial base. A healthy society is a productive society, while an unhealthy society will be an economically destitute society. Diseases that often ravage impoverished nations can curtail the productive capabilities of citizens. Recent global problems with contagious diseases, such as Bird Flu, SARS, Avian Flu, Enterovirus, and Ebola, have deleterious effects on the productive potentials of the nations affected. The destructive effects of pandemic and epidemic infectious diseases can be mitigated by prompt access to basic health care services. With fast global human interfaces nowadays, no nation can be immune to infectious diseases across the continents.

RELIABLE POWER SUPPLY

Electricity is a major source of power for production facilities. This fact has been known by developing countries for a long time. Yet not enough effort has been directed at adequate generation and reliable distribution of this important resource. Where electricity has been generated in abundant quantities, reliable distribution has been miserably lacking. Plants have been shortsightedly constructed without planning for adequate power supply. It's no wonder that most industries in developing economies are running only at a fraction of their capacities. The problems plaguing power supply companies should be critically studied so that a lasting solution can be found. A large portion of the initial development efforts should be directed at ensuring adequate power supply. Once supply is found to be reliable and stable, major production endeavors can then be pursued.

CONSISTENT WATER SUPPLY

In today's technology, chemicals play a significant role in product development. Water is an indispensable component of the use of chemicals in industrial operations. Water is needed not only to sustain life but also to support many products that make life livable. Water facility development should be addressed in the two categories below.

- Industrial water.
- Potable water.

Due to its huge volume of demand, industry should not compete with households for water supply. Like electricity, a reliable water supply should be assured before urging investment in production facilities. There are cases where a new industry opens, has water supply for a short period, and then is disappointed by the reality of water shortages.

ACCESSIBLE TRANSPORTATION SYSTEM

In agreement with Newton's Law of Motion, nothing moves without transportation. In the ancient days of geographically limited commerce, rapid transportation, though important, was less of a concern. In modern society, the global market has necessitated interactions with far-away locations. Developed economies appreciate the necessity of these interactions and they developed transportation systems to meet these needs. Developing economies are "still developing" because their transportation systems, in many respects, have yet to catch up with the rest of the world. Progress in modern commerce depends on the facilities for conveying products from one location to another efficiently, reliably, and safely.

After power and water, transportation should have the third priority in any strategic development plan for manufacturing development. Good roads should be constructed and maintained to provide a lasting support for development efforts. Industrial access roads as well as workforce access roads should be constructed. The economic loss can be enormous if the transportation system is neglected even in a developed economy. For example, the American Society of Civil Engineers (ASCE) reported that 42 percent of America's

Major urban highways remain congested, costing the economy an estimated $101 billion in wasted time and fuel annually (ASCE, 2013). The report confirms further that the USA Federal Highway Administration estimates that $170 billion in capital investment would be needed on an annual basis to significantly improve conditions and performance of the infrastructure. This is a huge financial requirement even for a wealthy nation. For developing nations, this book suggests incremental development, sustainment, and investment in transportation infrastructure. Little blocks of buildings are more manageable and sustainable. The development of a good transportation system can take a very long time at a very high cost. With its accessibility, the USA interstate system has fueled development in several parts of the USA.

EFFICIENT COMMUNICATION SYSTEM

Communication propels commerce. Telephone, Internet, and the postal service constitute alternate means of conducting business. In a developing economy, the lack of a reliable telephone service or reliable online systems can force entrepreneurs to make physical appearances when conducting even the most basic of business transactions. This, of course, necessitates the increased use of roads, thereby overloading the already inadequate transportation system. In an underdeveloped community, business that can be conveniently conducted over the telephone is normally conducted by physical presence. The requirement for physical mobility directly takes a person from productive activities, consequently creating adverse effects on national productivity.

In addition, frequent road transportation increases the exposure of the traveler to the hazards of the precarious transportation system. The more people need to be on the roads, the higher the congestion, and the greater the risks of road mishaps, which have been known to rob a nation of some of its most productive citizens. The reliability of electric power, water, transportation, and communication services is essential for a sound industrial development. Nations with well-developed communication systems have used them to further their development programs. On the other hand, those who have not paid proper attention to their communication systems will continue to struggle with dormant or regressing economies.

INDIGENOUS ECONOMIC MODELS

The formulation of an indigenous economic model is relevant for industrial development. The unique aspects of a developing economy that will necessitate indigenous models are discussed in this section. The reasons why imported economic principles and models may not work in a developing nation are presented. The success of indigenous economic models in an underdeveloped area can serve as a useful paradigm for industrial development endeavors in nations with similar economic plights. A synergistic implementation of indigenous economic models will have a positive impact on the overall world economy. Both the economic and non-economic aspects of industrial development must be accounted for in developing an indigenous model. Well-founded laws and principles of economics have been utilized in the process of formulating existing economic policies. Unfortunately, none of these policies has yielded totally satisfactory results in underdeveloped nations. There has been no short-term success, and there is no guarantee of a long-term success. Restrictions on importation, barter agreements, structural adjustment of currency value, and other economic policies are just a few examples of the concerted efforts being made by developing countries. The fact that some of these efforts have been or may be unsuccessful indicates that there are certain aspects of the domestic economic situation that are being overlooked, oversimplified, or unrecognized in these developing nations.

The standard economic models that are widely employed throughout the world were developed based on societal behaviors. Many emerging economies, with their peculiarities, simply don't fit the formulation of those models. That, perhaps, is why the economies in those societies have not responded positively to the traditional economic policies. Let us consider the laws of supply and demand, for example. These laws assume that consumers and producers will respond to certain inputs in some rational fashion. The downward-sloping demand curve indicates that consumers will buy less of a product as the price increases. By analogy, the upward-sloping supply curve indicates that suppliers will supply more of a product as the price increases. This is quite logical, except in certain underdeveloped economies. In many developing societies, the basic supply and demand laws may not be directly applicable. This is because people are expected to behave in manners foreign to their social, cultural, political, and economic structures and attitudes. Rather than try to mold a developing nation to fit existing economic models, attempts should be made to modify the models to fit the unique situations of the nations. Only then can we have models that accurately explain and predict the actions of "developing" consumers and, thus, provide reliable guidelines for effective national economic policies Boakye (2011) constructs a macroeconomic model of the Ghanaian economy. Ditto from Atta (1981). The model emphasizes the supply side of a small open economy. This presents a good example of an economic view tailored to an indigenous scenario. This is similar to what Badiru (1986) suggested for Nigeria.

The indigenous banking system that has been established in India is another excellent example of an economic strategy tailored to a unique national need. The indigenous banking system is based on the strength of personal relations, which is a strong social link in India. Personal relations are used to vouch for credit worthiness instead of using the conventional collateral approach. This indigenous banking system has worked very well, and it has made it possible for rural people, who would have otherwise gone to western-style banks, to actively engage in indigenous banking transactions. In a developing nation, the familiar curves of supply and demand may be mangled by several factors. Some of these factors are quite obvious. Some, on the other hand, are very subtle and can only be fully identified, understood, and quantified through dedicated economic research. Economic research must be directed specifically at formulating indigenous economic models. Some important factors that should be considered in such a research effort include the following.

- The level of propensity to consume in a developing nation (consumption orientation rather than production orientation).
- The inferiority complex that prompts some societies to prefer imported goods and services.
- The lack of self-pride in the products of local labor.
- The affinity for black-market transactions.

The factors above, in addition to others to be determined through appropriate locally focused research, should form the nucleus of an economic model for a developing country. As far back as 1975, the United Nations (UN) initiated a program to ensure that adequate quantitative information is available for the planning of economic and social development in Africa. Today, an indigenous economic model could be a significant component of similar quantitative modeling. It should be recognized that many developing nations can generate their own economic precedents. They typically have the number (in population) and the capability to support indigenous models. Through the practice of hoarding, "developing" retailers already know how to create artificial scarcity. They already know how to mount a social assault on the equilibrium point of the standard supply and demand curves. So, there is nothing like "market price" in existence in many developing nations. Some suppliers have already customized the principles of fair competition to accommodate collusion. Through the forces of collusion, the suppliers can collectively dictate their harsh terms to hapless consumers. To lay a serious foundation for industrial development, indigenous economists should be charged with the responsibility of conducting the appropriate research to study the feasibility of indigenous economic models for a nation that is aspiring to be industrialized. This may be effectively accomplished through post-graduate thesis research at the nation's institutions of higher learning. The research may be accomplished under the university-industry cooperative model suggested earlier.

In an in-depth analysis of national development planning, Gharajedaghi (1986) stressed the fact that a plan of action must not be shortsighted by descriptive adjectives that tell us nothing about the nature of the problem being solved. In government functions, there is a tendency to place responsibility for dealing with a problem in that part of government where a relevant functional adjective can be found. For example, if a problem is found in transportation, the blame is directed at the transportation department. The responsibility is automatically assumed to be that of the department even though the root of the problem may be a lack of public discipline. Adjectives and nouns used to describe problems (e.g., health, finance, social, economic, political, and so on) often tell us nothing about the real problem being faced. These adjectives simply indicate the point of view of the person looking at the problem. This pitfall must be avoided if a workable indigenous economic model is to be found. All aspects of a nation's economic system (e.g., social, education, information, etc.) must be assessed in the development of this model.

The benefits of having an indigenous economic model are numerous. If successful, it can encourage the country to undertake other self-help projects that might otherwise be left to external forces. It can serve as an incentive for citizens to believe in their own economic system. It can force people to exhibit more responsibility in their actions. The nation can exude pride for taking its own destiny in its hands. The implementation of the model can serve as an example and guideline for other countries in similar economic situations. The economic model can be presented to the citizens as an accurate description of their behaviors. That may compel them to reflect and become more conscious of the way they approach consumption of goods and services; thereby providing avenues for controlling inflation. The model can be an educational tool by making each sector of the economy more appreciative of the rationale behind the actions of other sectors. The idea of an indigenous economic model should provide the directions in which to look for solutions to pervasive development problems facing the nation. The development of a structure indigenous economic model will be subject to the following categories of factors.

Technology:

- Raw materials.
- Technology transfer.
- Local research.
- Skilled labor.
- Technology infusion.
- Local-context research.

Industry Diversity:

- Oil & Gas.
- Farming.
- Goods and services.
- Import-export transactions.
- Industrial production facilities.

External Influences:

- Foreign markets.
- Technology access.
- Trade obligations and restrictions.
- Laws and regulations.

Public Infrastructure:

- General public response.
- Social climate.
- Government agencies.
- Political system.
- Banking system.
- National productivity.
- Healthcare system.

Education:

- Educational system.
- Funding sources.
- Research engagements.
- Study abroad opportunities.
- Industry involvement.

Products:

- Preferences and options.
- Quality.
- Availability.
- Supply and demand.
- Cost and inflation.
- Brand options.

Entrepreneurial Opportunities:

- Business climate.
- Business loan.
- Financial resources.
- Taxes and incentives.
- Business etiquette.
- Collusion.
- Monopoly.

Government:

- Bribery and corruption.
- Investment climate.
- Exchange control.
- Immigration system.
- External Affairs.

Product standardization is another factor that can facilitate industrial development. If a standard is available, product interchangeability will be possible. Thus, the market for an industrial product can expand. With an expanded market, better levels of industrialization can be achieved. Units of measure have been one area where standardization has been pursued in many nations. For example, conversion to the metric system has been pursued by several nations to facilitate product compatibility and increase world trade. As more and more countries are switching to metric, a standard system of measurements will simplify international trade. The process of standardization is complicated, but once the initial difficulties of adopting a standard have been overcome, industrial exchanges will become easier. One of the major problems in industrial development is the lack of product integration. We may be effective in making good individual products, but when it comes to fitting the products together to arrive at an overall assembled system, we may have difficulties if there are no prevailing standards. Products should be designed with consideration for how they support one another to achieve an overall workable system in the match toward national industrialization.

REFERENCES

ASCE (2013), The American Society for Civil Engineers, 2013 Report Card for America's Infrastructure, (www.infrastructurereportcard.org), Reston, Virginia, 2013.

Atta, J. K (1981), **A Macroeconomic Mode of a Developing Economy**, Ghana, University Press of America, Accra, Ghana.1981.

Badiru, A. B. (1986), Indigenous Economic Model for Nigeria, ***National Concord newspaper***, Lagos, Nigeria, November 13, 1986, p. 2.

Badiru, A. B. (2012), **Project Management: Systems, Principles, and Applications**, Taylor & Francis CRC Press, Boca Raton, FL.

Badiru, A. B. (2016), **Global Manufacturing Technology Transfer: Africa-USA Strategies, Adaptations, and Management**, Taylor & Francis CRC Press, Boca Raton, FL.

Badiru, A. B., O. Ibidapo-Obe, and B. J. Ayeni (2019), **Manufacturing and Enterprise: An Integrated Systems Approach**, Taylor & Francis/CRC Press, Boca Raton, FL.

Boakye, S. (2011), "A Structural Macroeconomic Model of the Ghanaian Economy for Policy Scenario Analysis and Macroeconomic Forecasting," Technical Report, Ghana Ministry of Finance and Economic Planning, Accra, Ghana.

Gharajedaghi, J. (1986), **A Prologue to National Development Planning**, Greenwood Press, New York, NY.

NAE (2008), **National Academy of Engineering**, Grand Challenges for Engineering, Washington, DC.

4 A Model for Workforce-Driven Economic Development*

INTRODUCTION

Innovation for economic advancement is a critical part of workforce development. The work environment of today and the future requires an organizational workforce strategy that is innovative. The defense industry is predicated on having innovative workforce strategies that allow them to capitalize on multiple generations and subsets of the work-ready population. A community can make itself more marketable for economic development and industry re-location by promoting commitment to workforce innovation and a strong regional strategy. This chapter presents data analytics and strategies for workforce development and retention. The implicit and explicit connections between innovation and workforce development are detailed throughout the chapter.

This chapter came to life through four aerospace and defense industry studies and reports completed by the authors in Ohio in 2011, 2016, 2022, and 2024. In these studies, the authors investigated in-demand industries in Ohio and the need to focus on innovative workforce development to fully support the industries as they continue to grow in workforce demand. All of the previous studies have defined and highlighted details of the aerospace and defense industry while assessing the current and future growth of the industry. Building on their previous studies, the authors represent the aerospace and defense industry in Ohio with distinct clusters that best define the industry: aerospace manufacturing, Research and Development (R&D), federal and military, and aviation.

This chapter delves into the details of each distinct cluster specifically focusing on the workforce supply and demand. In addition, the aging workforce and demographic shifts are examined. This chapter highlights specific data analytics that can be completed in any industry sector to assist a region anywhere in the country in developing its next generation of workforce.

Data suggests that the higher education system in Ohio is not necessarily under-producing STEM graduates in the context of the patterns that support the defined clusters analyzed in this report. The analysis found Ohio's percentage of degree holders has experienced small, steady increases over the past decade. The rate of increase has been just above the U.S. average, but Ohio still lags behind the nation in the percentage of the population holding bachelor's degrees and graduate or professional degrees. Overall, the State's growth rate across demographics was comparable

* Adapted and modified from Barlow, Cassie, and Rochon (2020) and Schaeffer (2024)

DOI: 10.1201/9781003537755-4

with the overall population growth over the same period, a one-percent growth. This modest growth is directly tied to the largest industry sectors in the region. It may also be attributable to the economic development model employed across the state. Connecting economic growth to the growth of the regional military installation will lead to growth rates which are relative to the installation's growth.

There are workforce challenges in the aerospace and defense industry which cannot be ignored. While Ohio's colleges and universities are generally keeping pace with the required STEM workforce in R&D, aerospace manufacturing demands are outpacing training and education, which is further stressed by the replacement rates – retirements and other workforce churn. Nationally, more than 60 percent of the industry workforce supporting aerospace manufacturing is over 45 years old, with a median average age of 44.5. At the same time, only 21 percent of the workforce is under the age of 35, meaning that employee retirement will continue to drive higher workforce demand.

Aerospace manufacturing directly supports more than 25,000 jobs in Ohio, ranking it third in the country. Ohio's growth is moderate in this industry cluster in Ohio. Aerospace manufacturing in Northeast and Southwest Ohio are both projected to continue growing. The aerospace manufacturing workforce requires some of the highest-skilled manufacturing workers. Almost 20 percent of its workforce are engineers.

Ohio ranks 13th in the country in the number of R&D jobs, with a projected growth of 25 percent in the next decade. This cluster is expected to have significant projected growth. Economic growth can be realized through dedicated efforts to leverage this research for commercial applications. Analysis of R&D expenditure data suggests that Ohio has a significant emphasis on companies and individuals contracted to conduct this R&D compared to the top ten states in R&D expenditures. More analysis is necessary to understand why this is occurring in Ohio and who owns the intellectual property being produced because of this structure. Ohio ranks ninth in the U.S. for the number of government civilian jobs. Like most of the other states employing large numbers of civilian and military workers, Ohio's federal jobs are projected to continue growing. This aligns with federal budgets and expenditures.

To best address the workforce demand within the aerospace and defense industry in Southwest Ohio, the authors recommend building a robust ecosystem with participation across the region that includes tapping into all potential sources of workforce. This ecosystem requires a broad audience with maximum participation from all parts of the ecosystem. Strong leadership, with a substantive regional workforce strategy, is required and depends on participation from all of the stakeholders. The ecosystem will be successful if there is a high-level workforce strategic plan that is regularly monitored and measured.

This chapter can serve as a step-by-step model for any region of the country that desires to focus on the workforce and economic development of an in-demand industry.

The aerospace and defense industry is critical to national defense and therefore a very important and unique industry in the world market. It's also a very large industry with many complexities. The aerospace and defense industry serves both commercial and defense markets with a vast array of products. The industry also includes some very large employers, such as Boeing and Lockheed Martin, as well as thousands of smaller companies. The industry is smaller than it was but is still quite large and has substantive strategic importance in the U.S.

The end of the Cold War meant decreased spending in aerospace and defense, which led to diversification of the industry. There were many new commercial applications discovered for aerospace and defense products. This also led to the Department of Defense spending a much smaller percentage of their budget on research and development. With the conclusion of the Cold War, the U.S. went through a recession, which led to the industry cutting workforce and consolidating operations. Examples of these consolidations include Northrop Grumman and Lockheed Martin. Due to the recession and reduction in spending, aerospace and defense companies started to focus sales on foreign governments. This new strategy worked until the Asian economic crisis in the late 1990s. This crisis resulted in another hit to the market bottom line, which led to more reductions for companies across the industry. After 9/11, the industry was again impacted, but this time by a large infusion of new dollars that were part of an investment into a new conflict in the Middle East. The aerospace and defense industry is forecasted to continue growing through modernization and recapitalization.

Due to the complexity of the aerospace and defense industry and the vast nature of the market, the industry is the world's leading producer of technology and supports one of the largest high-skill and high-wage workforces in the country. In 2015, the U.S. aerospace and defense industry supported almost 2.1 million jobs (AIA, Dec 2021). The primary share of jobs within the aerospace and defense industry is in manufacturing (917,000 jobs) and then information and professional services (465,000) (AIA, Dec 2021). Growth areas within the industry include foreign sales as well as research and development. An evolving demand and continued growth within the aerospace and defense industry as well as an aging workforce will make the next 20 years within the industry a challenge. In this chapter, the authors focus on workforce demand and how a region can build a robust ecosystem to react to needs within an industry, as documented in Barlow and Rochon, 2020; Aerospace Industries Association 2023; Association of Unmanned Vehicle Systems International 2023; Boeing 2023; Bureau of Labor Statistics databases; National Science Foundation Science and Engineering Indicators State Data Tool; U.S. Census Bureau; Ziprecruiter 2023.

Defining the Industry

To thoroughly examine the supply and demand of workforce within the aerospace and defense industry, the authors chose to scope their work in terms of a region within Ohio, specifically, the Southwestern region. To represent the existing demand within the aerospace and defense industry, the authors divided the industry into four key areas:

- Aerospace Manufacturing.
- Research and Development (including Information Technology related to Aerospace).
- Federal and Military.
- Aviation.

TABLE 4.1
Aerospace Manufacturing Occupations

Code	Description
333314	Optical Instrument and Lens Manufacturing
334511	Search, Detection, Navigation, Guidance, Aeronautical, and Nautical System and Instrument Manufacturing
334515	Instrument Manufacturing for Measuring and Testing Electricity and Electrical Signals
335931	Current-carrying Wiring Device Manufacturing
336411	Aircraft Manufacturing
336412	Aircraft Engine and Engine Parts Manufacturing
336413	Other Aircraft Parts and Auxiliary Equipment Manufacturing
336414	Guided Missile and Space Vehicle Manufacturing
336415	Guided Missile and Space Vehicle Propulsion Unit and Propulsion Unit Parts Manufacturing
336419	Other Guided Missile and Space Vehicle Parts and Auxiliary Equipment Manufacturing

Aerospace Manufacturing

Although aircraft are not manufactured in Ohio, many aerospace manufacturing suppliers are located in Ohio. For instance, GE Aviation manufactures jet engines and other components in Southwest Ohio. In 2024, Boeing reported that it had 306 suppliers and vendors in Ohio, accounting for $7B in total economic impact that supported up to 385,000 jobs in Ohio (Boeing, 2024).

Aerospace manufacturing is an important component of Ohio's economy. Ohio supports direct aerospace manufacturing jobs defined by the cluster in Table 4.1. However, this cluster does not define all aerospace manufacturing since occupations such as metalworking, sensors, or composites may serve automotive, medical, or other kinds of manufacturing in addition to the aerospace industry. The direct aerospace manufacturing jobs are defined with the following industry codes.

Research and Development (R&D)

The Aerospace R&D cluster includes more than just research companies but also includes testing, engineering services, software, and computer design, including the following occupational codes in Table 4.2.

Federal/Military

Federal jobs are a substantial part of Ohio's aerospace industry. Wright-Patterson Air Force Base (WPAFB) in Dayton, Ohio, is the largest, single-site employer in the state, employing more than 38,000 civilians and military personnel. Ohio is also home to the sixth-largest National Guard in the U.S. Defense Finance and Accounting Service, and Defense Logistics Agency. NASA Glenn in Cleveland is an important part of Ohio's aerospace industry and is a large federal employer, too. See Table 4.3.

TABLE 4.2
Research and Development Occupations

Code	Description
511210	Software Publishers
541330	Engineering Services
541380	Testing Laboratories
541512	Computer Systems Design Services
541690	Other Scientific and Technical Consulting Services
541712	Research and Development in the Physical, Engineering, and Life Sciences (Except Biotechnology)

TABLE 4.3
Federal/Military Occupations

Code	Description
901199	Federal Government, Civilian, Excluding Postal Service
901200	Federal Government, Military

TABLE 4.4
Aviation Occupations

Code	Description
481111	Scheduled Passenger Air Transportation
481112	Scheduled Freight Air Transportation
481211	Nonscheduled Chartered Passenger Air Transportation
481212	Nonscheduled Chartered Freight Air Transportation
481219	Other Nonscheduled Air Transportation
488111	Air Traffic Control
488119	Other Airport Operations
488190	Other Support Activities for Air Transportation
611512	Flight Training

AVIATION

While aviation is directly linked to aerospace, from a workforce development perspective, it is a demand that follows the market, not a business need. As consumers spend more on commercial flights and as businesses become more nationally and internationally connected, the demand for pilots and other support staff will naturally grow. As this happens, commercial airlines will look for a pool of qualified pilots. That search typically starts with Department of Defense pilots and for this reason, the growth in commercial pilots is worth benchmarking. See Table 4.4.

National Workforce Trends

As shown in Figure 4.1, the aging workforce in the U.S. has been a popular topic among economists and the government for decades. There is good reason for this concern and discussion of this topic. The aging baby boomers, recession of the late 2000s, the COVID-19 pandemic, as well as a rising life expectancy, have changed the face of our current and future workforce. Internationally, the number of older people is expected to exceed the number of children for the first time by 2047. Although the retirement age seems to be increasing, there is still a need to replace this generation with a new workforce as they retire. The companies that figure out how best to use

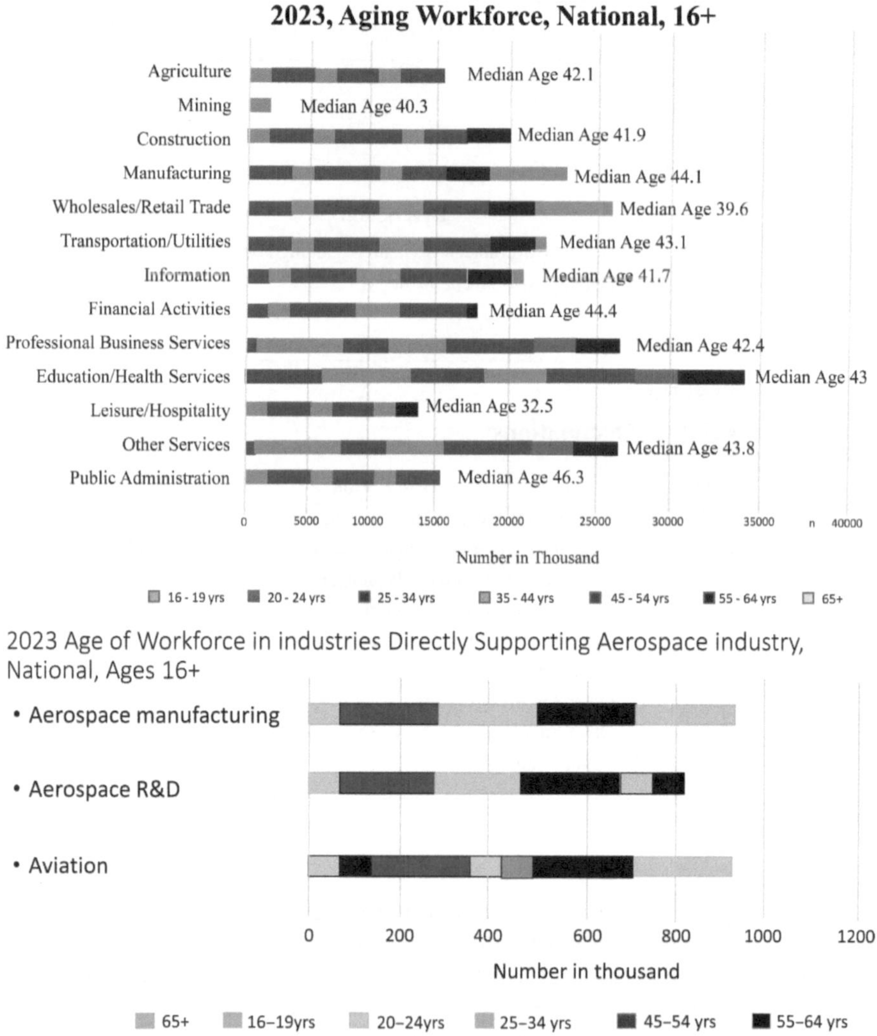

FIGURE 4.1 National workforce population age statistics.

(Source: Bureau of Labor and Statistics from the Current Population Survey, 2023)

this aging workforce to their betterment, for example, in training and mentoring of the new workforce or as a part-time workforce, will be the most successful companies in the world. This generation has extensive knowledge and skills that continue to contribute to companies across the country. At the same time, strategic workforce planning is critical as baby boomers leave the workforce.

Nationally, more than half of the workforce in the industries supporting aerospace manufacturing is over 45 years old with only a quarter below the age of 30. This composition is slightly better for the R&D industry with 43 percent over the age of 45 and 32 percent below the age of 35. This is a metric to watch as the workforce between 35 and 44 years old is critical for the long-term growth and stability of the industry. Industry experience has been a common theme in aerospace and defense workforce shortage studies. The higher education system can develop strategies to increase competition rates in desired degrees; however, industry must take responsibility for developing their workforce to meet future experience demands.

In addition to an aging workforce, there are other demographic changes occurring in the workforce that must be part of an organizational strategic workforce plan. The gender balance in the U.S. has been tipped towards women for the last decade and is going to continue in this direction. In the year 2050, there will be 150 women for every 100 men in the U.S. In addition, there has been a continual increase in African American and Hispanic populations. A successful organization will figure out how to recruit, hire, and retain a diverse workforce.

Current State of the Industry

With the clusters defined, a deeper dive has provided some unexpected projections. In this section, the authors use a modeling tool that considers demand based on federal data sources that project supply and demand using algorithms applied to historical trends, the supply of workers completing higher education programs that meet those demands, and churn. Included in this data is the labor market information data that 'Ohio Means Jobs' uses to identify the most in-demand jobs. By layering this data set on top of the supply data and churn data, a more detailed picture develops. The bottom line in many of these tables is that Ohio is not necessarily underproducing STEM graduates in the context of the staffing patterns that support the defined industry clusters outlined.

STEM Talent Analysis

To validate this conclusion, National Science Foundation data was studied to understand if this was a projection unique to the staffing patterns evaluated or true of the broader STEM workforce. See Figure 4.2.

The data indicates that the number of Science and Engineering degrees conferred in Ohio increased by 71 percent over ten years, and the percentage of Science and Engineering Degrees compared to all higher education degrees conferred has increased slightly in the past few years getting closer to the national trend. At the same time, occupations requiring those degrees have seen moderate increases (see Figure 4.3).

This data suggests that the state is seeing an increase in STEM degrees, which are meeting the increasing workforce demand, and is likely generating a surplus of Science and Engineering degrees conferred. It is important to note that prior to the

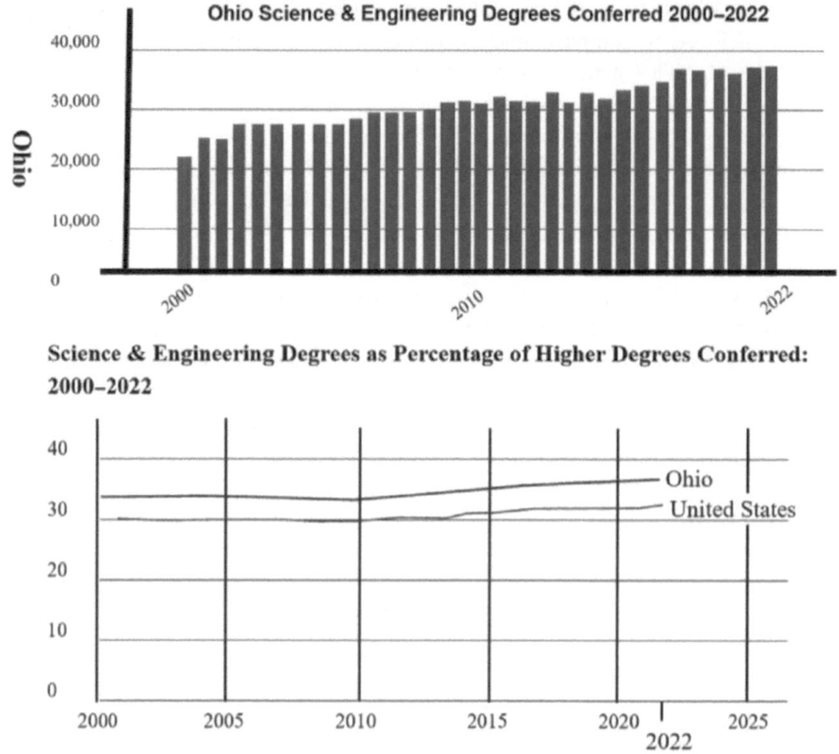

FIGURE 4.2 Changes in science and engineering degrees, 2000–2022.

(Source: NSF, 2021 Science and Engineering Indicators State Data Tool)

recession, Ohio employed more engineers than the national average, but the loss of large tech companies, such as Delphi and NCR, greatly impacted the overall engineering occupations. Data shows the state once again reached 2003 ratios in 2012 and has continued to increase its engineers as a percentage of all occupations, surpassing the national average. The loss of commercial positions led to a large shift in available engineers seeking jobs while the DoD work continued to grow both on and off base. The bottom line is that STEM degrees continue to increase as well, in terms of percentage of total degrees. Ohio's growth in these areas is right in line with national trends and in terms of percentage of all occupations has surpassed national trends. There is continued work required to sustain the output of STEM majors to meet the demand of the regional industries focused on market trends and consumer demand. As the staffing patterns in the clusters identified in this report are compared to the supply, it will be important to keep in mind the overall STEM workforce supply.

Concerning the aerospace manufacturing, R&D, Federal/Military, and aviation clusters, the talent supply appears to be sufficient to meet workforce demand. Many factors are coming into play: 1) concerted efforts by a robust network of STEM initiatives are making headway and addressing the STEM talent supply gap, 2) there are major gaps in skilled jobs that do not require higher education and are thus harder

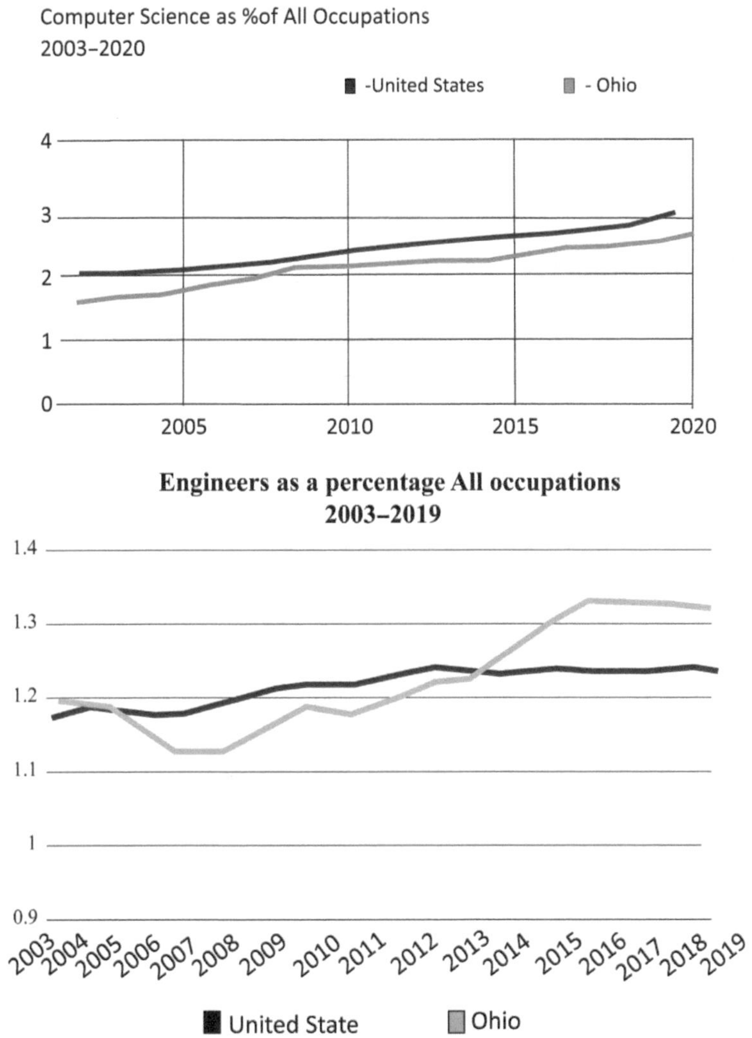

FIGURE 4.3 Changes in engineers and computer specialists, 2003–2020.

(Source: NSF, 2021 Science and Engineering Indicators State Data Tool)

to measure supply statistics (e.g. Team Assemblers), 3) the clusters with growth projected to meet or exceed national growth rates are the R&D and aerospace manufacturing clusters 4) completion rates are up, but retention of students in the state of Ohio is moderate and brain drain is impacting the shortages that have been reported by businesses, and 5) this data is based on historical trends from secondary data sources only, so it's possible there are STEM-related workforce shortages that could exist due to niche sectors inaccurately portrayed by the available data. Regarding factors three and four listed above, a recommendation section has been produced that highlights the need for regional workforce strategies.

Talent Retention

One item that is often discussed when reviewing degrees conferred is the retention of talent (Figure 4.4). "Brain Drain" is a concern, especially when considering in-demand STEM degrees. A review of American Community Survey data from the U.S. Census Bureau can provide insight into degree holders in Ohio and within each Jobs Ohio region. Ohio's percent of degree holders has experienced small, steady increases over the past six years according to the census five-year trend data. The rate of increase has been just above the U.S. average, but Ohio still lags the nation in the percentage of the population holding bachelor's degrees and graduate or professional degrees. The State's growth rate in these demographics was comparable with the overall population growth over the same period, a one percent growth. The national growth rate, however, was six percent while the growth in degree holders was all less than one percent (see Figure 4.5).

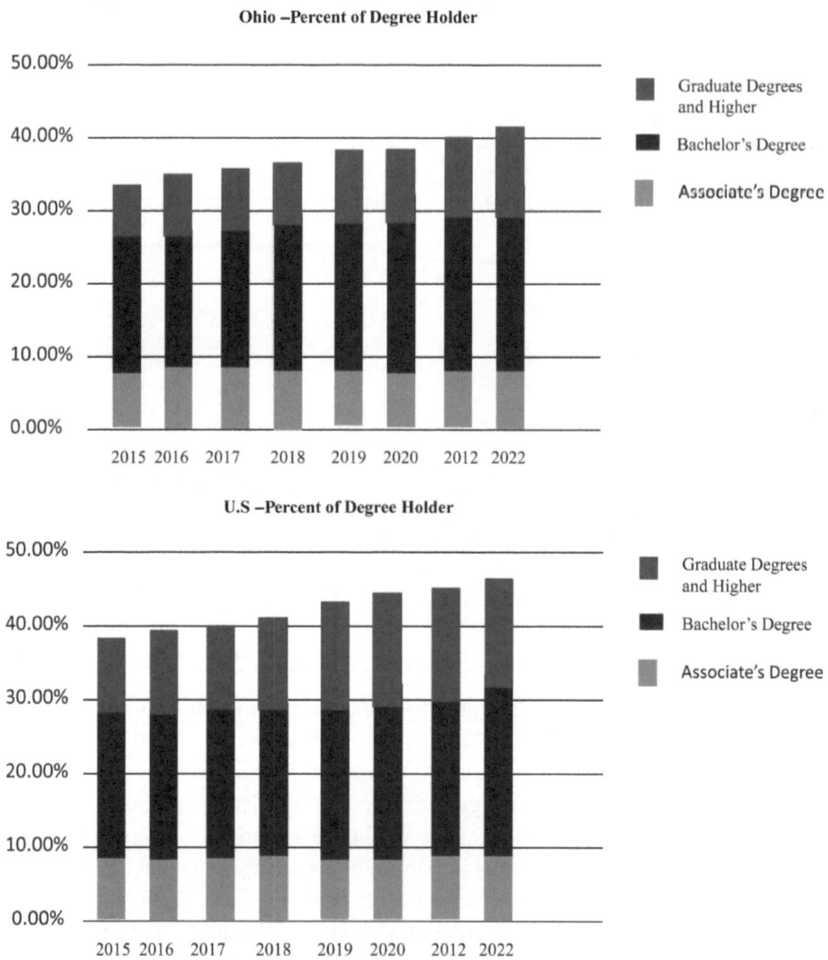

FIGURE 4.4 Percent of degree holders in Ohio and the U.S., 25+ years old, 2015–2022. **(Source: American Community Survey, U.S. Census Bureau, 5-year estimates)**

Percentage of Degree Holders – JobsOhio Region – Western

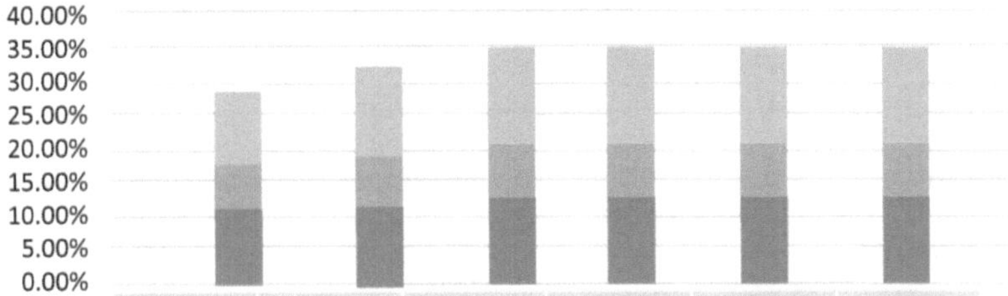

	2015	2016	2017	2018	2019	2020
Graduate Degree and Higher	9.28%	9.41%	9.85%	10.06%	10.32%	10.60%
Bachelor Degree	13.80%	14.20%	14.38%	14.71%	14.98%	10.52%
Associate Degree	9.17%	9.35%	9.57%	9.54%	9.52%	9.65%

Graduate Degree and Higher   Bachelor Degree   Associate Degree

Percentage of Degree Holders – JobsOhio Region – Central

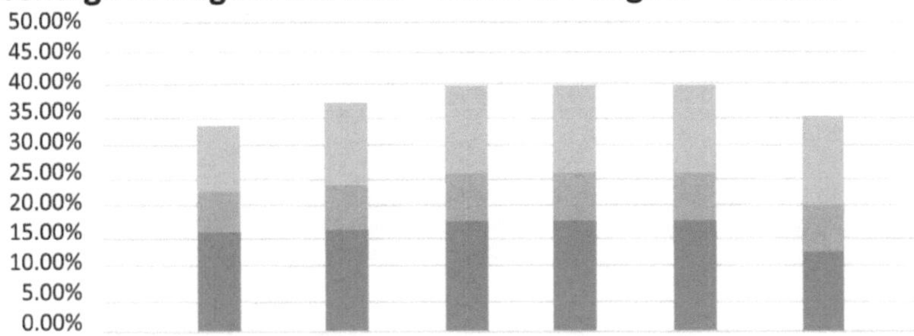

	2015	2016	2017	2018	2019	2020
Graduate Degree and Higher	11.99%	12.18%	12.44%	12.77%	12.99%	11.29%
Bachelor Degree	21.41%	21.87%	22.17%	22.53%	22.86%	23.23%
Associate Degree	7.17%	7.21%	7.30%	7.42%	7.44%	7.52%

Graduate Degree and Higher   Bachelor Degree   Associate Degree

Percentage of Degree Holders – JobsOhio Region – Southwest

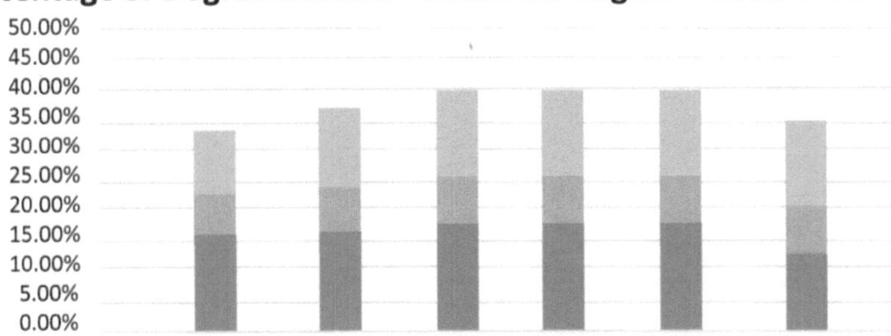

	2015	2016	2017	2018	2019	2020
Graduate Degree and Higher	12.05%	12.47%	12.70%	11.10%	11.37%	11.77%
Bachelor Degree	20.51%	21.84%	22.24%	21.39%	21.80%	21.19%
Associate Degree	8.17%	8.19%	8.41%	8.45%	8.43%	8.53%

Graduate Degree and Higher   Bachelor Degree   Associate Degree

FIGURE 4.5 Percent of degree holders by JobsOhio regions, 25+ years old, 2015–2020. *(Continued)*

(Source: American Community Survey, U.S. Census Bureau, 5-year estimates)

Percentage of Degree Holders – JobsOhio Region – Northwest

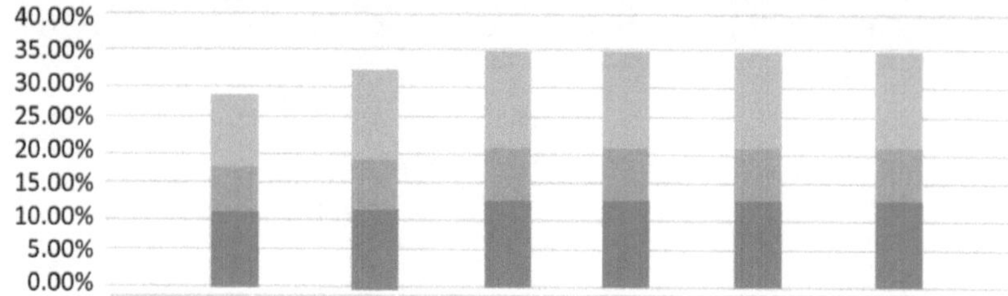

	2015	2016	2017	2018	2019	2020
Graduate Degree and Higher	8.09%	8.17%	8.43%	8.62%	6.66%	8.85%
Bachelor Degree	12.88%	13.23%	13.57%	13.81%	14.16%	14.47%
Associate Degree	9.73%	10.08%	10.14%	10.20%	10.53%	10.53%

Graduate Degree and Higher Bachelor Degree Associate Degree

Percentage of Degree Holders – JobsOhio Region – Southeast

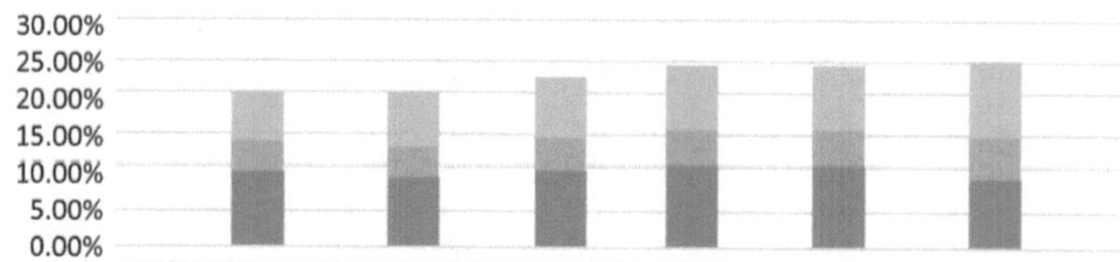

	2015	2016	2017	2018	2019	2020
Graduate Degree and Higher	5.55%	5.62%	5.88%	5.94%	5.99%	6.10%
Bachelor Degree	8.90%	9.22%	9.28%	9.36%	9.52%	9.69%
Associate Degree	8.98 %	9.29%	9.33%	9.24%	9.44%	9.63%

Graduate Degree and Higher Bachelor Degree Associate Degree

Percentage of Degree Holders – JobsOhio Region – Northeast

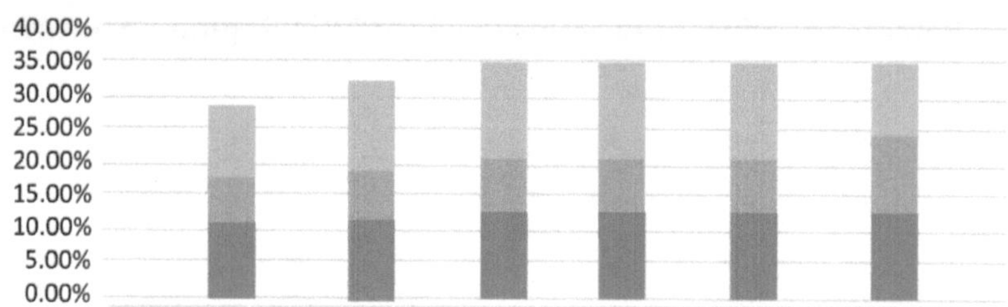

	2015	2016	2017	2018	2019	2020
Graduate Degree and Higher	9.48%	9.72%	9.92%	10.20%	10.56%	10.60%
Bachelor Degree	16.02%	16.27%	16.55%	16.75%	16.99%	17.27%
Associate Degree	7.84%	8.05%	8.19%	8.29%	8.44%	8.62%

Graduate Degree and Higher Bachelor Degree Associate Degree

FIGURE 4.5 *(Continued)*

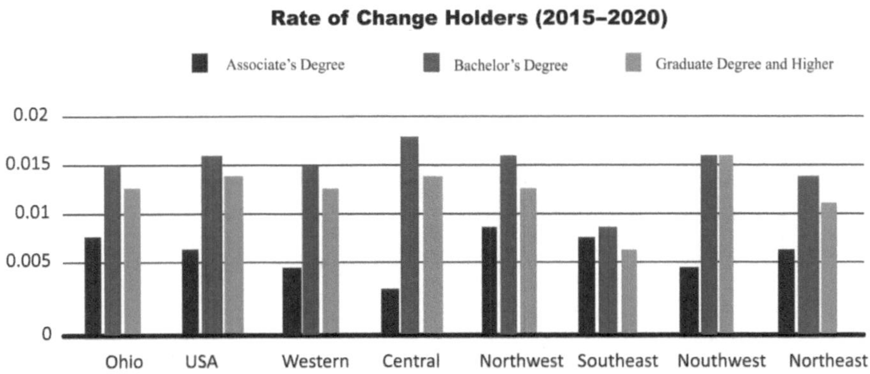

FIGURE 4.6 Rate of change of education attainment by region, 25+ year olds, 2015–2020. **(Source: American Community Survey, U.S. Census Bureau, 5-year estimates)**

When considering the challenges reported by industry to find qualified workers, a closer look was warranted. When examining the JobsOhio regions, a positive trend was evident. The Central region leads the state in the percentage of degree holders, followed closely by the Southwest region. Each region experienced a positive trend in degree holders.

This trend continues when analyzing 24- to 35-year-olds in each JobsOhio region. This age group is often considered the most critical retention demographic. Connecting graduates to employment in Ohio is critical in retaining the needed talent. Despite a minor dip in the West region in 2013, each region experienced modest growth each year analyzed. In addition, for each region except for the Northeast region, the overall population of 25- to 34-year-olds decreased. This suggests that the major migration out of each region other than the Northeast is not predominantly bachelor's degree holders. This data led to questions that could be further explored in each region. More research must be done to determine if the reported talent gap by businesses experiencing hiring difficulties is due to a mismatch in degrees conferred compared to industry need or if there is a disconnect between jobseekers and businesses seeking talent. The rate of change in each region could serve as a metric to identify pockets of growth or areas that are improving because of initiatives to retain young talent (see Figure 4.6).

Aerospace Manufacturing

Aerospace manufacturing directly supports more than 25,000 jobs in Ohio, ranking it eighth in the country. Growth predictions for aerospace manufacturing in Ohio for the next decade are modest.

The location quotient, (occupation percentage of local employment/occupation percentage of national employment), permits a comparison of the relative importance of the occupation in the regional employment mix relative to the country (see Table 4.5).

TABLE 4.5
Top States for Aerospace Manufacturing Jobs

State	State Name	2020 Jobs	2030 Jobs	2020–2030 Change	2020–2030 % Change	2020 Location Quotient
6	California	128,042	112,574	(15,468)	(12)	1.42
53	Washington	92,904	88,077	(4,827)	(5)	5.17
48	Texas	59,744	60,136	392	1	0.92
12	Florida	40,911	45,206	4,295	10	0.90
4	Arizona	37,135	36,770	(365)	(1)	2.49
9	Connecticut	35,113	34,078	(1,035)	(3)	4.12
20	Kansas	34,459	30,274	(4,185)	(12)	4.71
39	Ohio	25,072	25,992	850	3	0.90
13	Georgia	22,003	21,126	(877)	(4)	0.94
36	New York	20,280	19,542	(738)	(4)	0.42
25	Massachusetts	20,237	17,760	(2,477)	(12)	1.08
29	Missouri	19,531	20,631	1,100	6	1.33
42	Pennsylvania	18,257	18,101	(156)	(1)	0.60
24	Maryland	15,162	16,395	1,233	8	1.08
1	Alabama	14,864	15,811	947	6	1.43
40	Oklahoma	14,596	17,615	3,019	21	1.73
8	Colorado	14,360	14,649	289	2	1.01
49	Utah	12,268	13,459	1,191	10	1.54
33	New Hampshire	11,885	12,749	864	7	3.51
37	North Carolina	11,797	11,763	(34)	(0)	0.50

Source: EMSI, 2020 – QCEW Employees, Non-QCEW Employees, and Self-Employed

Although Ohio's growth is modest in aerospace manufacturing, Ohio's metropolitan statistical areas (MSAs) vary. For the most part, aerospace manufacturing in Northeast Ohio is projected to decline, while in Southwest Ohio it's predicted to grow. Vibrancy and growth of the aerospace manufacturing industry earned the Southwest Ohio Aerospace Region (SOAR) a manufacturing designation from the Economic Development Administration's Investing in Manufacturing Communities Partnership (IMCP) initiative. Together, the Dayton and Cincinnati metro areas provide more than 16,000 jobs, representing two-thirds of Ohio's aerospace manufacturing industry. See Table 4.6.

The aerospace manufacturing workforce requires some of the highest-skilled manufacturing workers. Almost 20 percent of its workforce are engineers – aerospace, industrial, mechanical, electrical, and others. Also required are machinists and other highly-skilled production workers, technicians, software publishers, and managers at various levels.

The following table shows the top jobs in the industry, making up just over 50 percent of all jobs in aerospace manufacturing. Since the growth of the statewide aerospace manufacturing industry is modest, the occupational demands are not dramatic. Although industrial engineers and software developers show healthy growth, the overall job growth remains relatively modest – thirty-four positions over ten years. See Table 4.7.

TABLE 4.6
Ohio Aerospace Manufacturing Jobs by MSA

MSA Name	2020 Jobs	2030 Jobs	2020–2030 Change	2020–2030 % Change	2020 Location Quotient
Cincinnati, OH-KY-IN	14,660	14,771	111	1	2.69
Cleveland-Elyria, OH	4,794	4,643	(151)	(3)	0.92
Dayton-Kettering, OH	2,790	3,795	1,005	36	1.45
Akron, OH	1,388	1,380	(8)	(1)	0.84
Toledo, OH	631	788	157	25	0.40
Mansfield, OH	369	253	(116)	(31)	1.45
Columbus, OH	334	293	(41)	(12)	0.06
Lima, OH	255	344	(89)	35	1.00
Springfield, OH	144	113	(31)	(22)	0.59
Portsmouth, OH	123	191	68	55	1.02

Source: EMSI, 2020 – QCEW Employees, Non-QCEW Employees, and Self-Employed.

TABLE 4.7
Top Jobs in the Aerospace Manufacturing Industry in Ohio

SOC	Description	Employed in Industry Group (2020)	Employed in Industry Group (2030)	Change (2020–2030)	% Change (2020–2030)	% of Total Jobs in Industry Group (2020)	Median Hourly Earnings
17-2011	Aerospace Engineers	1,424	1,521	97	7	5.7%	$55.27
17-2112	Industrial Engineers	1,400	1,577	177	13	5.6	$38.77
51-9061	Inspectors, Testers, Sorters, Samplers, and Weighers	1,216	1,082	(134)	(11)	4.8	$18.97
15-1256	Software Developers and Software Quality Assurance Analysts and Testers	1,205	1,323	118	10	4.8	$44.29
51-4041	Machinists	1,013	1,063	50	5	4.0	$20.46
51-2098	Miscellaneous Assemblers and Fabricators	915	802	(113)	(5)	3.7	$16.40
51-2028	Electrical, Electronic, and Electromechanical Assemblers, Except Coil Winder, Tapers, and Finishers	873	900	27	7	3.5	$16.64

(Continued)

TABLE 4.7 *(Continued)*
Top Jobs in the Aerospace Manufacturing Industry in Ohio

SOC	Description	Employed in Industry Group (2020)	Employed in Industry Group (2030)	Change (2020–2030)	% Change (2020–2030)	% of Total Jobs in Industry Group (2020)	Median Hourly Earnings
17-2141	Mechanical Engineers	838	881	43	5	3.3	$38.23
51-9161	Computer Numerically Controlled Tool Operators	646	614	(32)	6	2.6	$19.17
49-3011	Aircraft Mechanics and Service Technicians	565	604	39	5	2.3	$29.49
43-5061	Production, Planning, and Expediting Clerks	552	582	30	5	2.2	$22.92
17-2199	Engineers, All Other	551	586	35	6	2.2	$43.64
51-1011	First-Line Supervisors of Production and Operating Workers	538	565	27	5	2.1	$29.27
17-2071	Electrical Engineers	511	545	34	7	2.0	$40.78
13-1028	Buyers and Purchasing Agents	488	484	(4)	(1)	1.9	$30.12
13-1198	Project Management Specialists and Business Operations Specialists, All other	473	509	36	8	1.9	$33.81
51-2011	Aircraft Structure, Surfaces, Rigging, and System Assemblers	403	400	(3)	(1)	1.6	$22.81
51-4081	Multiple Machine Tool Setters, Operators and Tenders, Metal and Plastic	402	413	11	3	1.6	$17.15
17-3021	Aerospace Engineering and Operations Technologist and Technicians	371	403	32	9	1.5	$32.70
13-1081	Logisticians	370	404	34	9	1.5	$34.84

Source: EMSI, 2020 – QCEW Employees, Non-QCEW Employees, and Self-Employed.

However, looking at the overall occupational demand for those positions across all industries shows a much different picture. Ohio's aging workforce, replacements due to retirement, and other churn will impact many occupations, particularly within the manufacturing industry. Regardless of whether manufacturing for the auto or aerospace industry, higher-skilled positions are required in all manufacturing, increasing the demand for those skills. See Table 4.8.

TABLE 4.8
Gap Analysis of Top Aerospace Manufacturing Jobs Across All Industries

Description	2021 Jobs	2030 Jobs	2021–2030 % Change	2021–2030 Change	Avg. Annual Openings	Regional Completions	Gap
Aerospace Engineering and Operations Technologists and Technicians	945	1,007	7	62	89	210	121
Aerospace Engineers	3,894	4,057	4	162	244	322	78
Aircraft Mechanics and Service Technicians	5,424	5,654	4	230	453	258	(195)
Aircraft Structure, Surfaces, Rigging, and Systems Assemblers	484	477	(1)	(7)	47	109	62
Architectural and Engineering Managers	6,214	6,475	4	261	419	9,788	9,369
Buyers and Purchasing Agents	18,322	17,576	(4)	(746)	1,629	406	(1,223)
Computer Numerically Controlled Tool Operators	12,274	11,563	(6)	(711)	1,053	51	(1,002)
Computer Systems Analysts	27,890	29,211	5	1,321	1,952	2,954	1002
Electrical, Electronic, and Electromechanical Assemblers	11,769	12,432	6	663	1,267	438	(829)
Electrical Engineers	7,109	7,469	5	360	479	1,349	870
Engineers, All other	8,224	8,478	3	254	532	1,040	508
First-Line Supervisors of Production and Operating Workers	32,323	32,800	1	477	2,999	630	(2,369)
Industrial Engineers	15,635	16,994	9	1,359	1,118	632	(486)
Industrial Machinery Mechanics	21,860	24,633	13	2,773	2,175	165	(2,010)

(Continued)

TABLE 4.8 ***(Continued)***
Gap Analysis of Top Aerospace Manufacturing Jobs Across All Industries

Description	2021 Jobs	2030 Jobs	2021–2030 % Change	2021–2030 Change	Avg. Annual Openings	Regional Completions	Gap
Industrial Production Managers	11,746	11,871	1	125	767	12,363	11,596
Inspectors, Testers, Sorters, Samplers, and Weighers	30,987	27,383	(12)	(3,604)	3,160	7	(3,153)
Logisticians	8,262	8,811	7	550	778	1,284	506
Machinist	25,763	26,390	2	627	2,544	88	(2,456)
Mechanical Engineers	15,465	16,094	4	628	975	2,251	1,276
Miscellaneous Assemblers and Fabricators	78,582	71,639	(9)	(6,943)	7,207	0	(7,207)
Multiple Machine Tool Setters, Operators and Tenders Metal and Plastic	15,834	15,808	(0)	(27)	1,487	88	(1,399)
Production, Planning, and Expediting Clerks	20,154	20,678	3	523	1,985	0	(1,985)
Project Management Specialists and Business Operations Specialist	47,080	49,421	5	2,341	4,428	171	(4,257)
Shipping Receiving and Inventory Clerks	38,830	36,615	(6)	(2,214)	3,414	13	(3,401)
Software Developers and Software Quality Assurance Analysts	47,015	54,261	15	7,246	3,890	3,303	(587)
	512,086	517,796	1	5,711	45,093		

Source: EMSI, 2020 – QCEW Employees, Non-QCEW Employees, and Self-Employed

While many of the production occupations do not necessarily require postsecondary training, they still require on-the-job training. Manufacturers across Ohio have expressed difficulty attracting a skilled workforce. These projections suggest that demand will only increase and could potentially have negative effects on Ohio's aerospace manufacturing.

Research and Development

Ohio's R&D market has strong growth opportunities in the aerospace and defense industry. With many research universities across the State, the Air Force Research Laboratory (AFRL), NASA Glenn, and a large concentration of R&D companies doing Small

TABLE 4.9
Top States for Research and Development Jobs

State Name	2020 Jobs	2030 Jobs	2020–2030 Change	2020–2030 %Change	2020 Location Quotient
California	539,539	598,869	59,330	11	1.26
Texas	297,810	367,471	69,661	23	0.98
Virgina	197,812	215,112	17,300	9	2.09
New York	167,860	195,531	27,671	16	0.76
Florida	165,559	195,075	29,516	18	0.78
Washington	152,308	182,569	30,261	20	1.83
Massachusetts	143,343	158,338	14,995	10	1.67
Michigan	136,375	137,490	1,115	1	1.35
Pennsylvania	129,374	147,033	17,659	14	0.93
Maryland	120,452	146,335	25,883	21	1.87
Colorado	112,741	129,675	16,934	15	1.70
Illinois	106,867	113,231	6,364	6	0.77
North Carolina	100,575	121,851	21,276	21	0.93
Georgia	98,496	116,397	17,901	18	0.91
Ohio	94,205	106,566	12,361	13	0.74
New jersey	93,780	98,517	4,737	5	0.99
Missouri	57,862	70,906	13,044	23	0.86
Alabama	53,500	63,018	9,518	18	1.13
Minnesota	52,926	58,988	6,062	11	0.78
Arizona	52,557	66,080	13,523	26	0.76

Source: EMSI, 2020 – QCEW Employees, Non-QCEW Employees, and Self-Employed.

Business Innovation Research (SBIR) to support federal agencies, Ohio has a strong R&D foundation, which is projected to grow even stronger within the next decade. Although the R&D concentration is lower than the national average, Ohio ranks 15th in the number of R&D jobs, with a projected growth of 25 percent in the next decade.

Organizations such as Brookings have reported that more than 90 percent of a community's job growth is organic (Table 4.9). Less than ten percent will come from recruiting businesses from other communities. With this being the case, investing in the recruitment of major aerospace manufacturers and even an aircraft manufacturer could be viewed as a flawed strategy. This becomes clearer once one considers the possibility for organic growth to occur from a healthy, growing research and development sector. Investing in the spin out of new ventures from innovations in aerospace is the state's best chance at turning the aerospace manufacturing industry around and into strong growth numbers. This strategy is a long-term strategy as the job numbers will likely follow a logarithmic or exponential curve, meaning initial investments typically won't result in headline-worthy job numbers, but over time will produce greater results than incentive programs for business relocation.

Like Aerospace Manufacturing, the Aerospace R&D industry is also more heavily concentrated in Southwestern Ohio, especially in the Dayton metropolitan area,

TABLE 4.10
Ohio R&D Jobs by MSA

MSA Name	2020 Jobs	2030 Jobs	2020–2030 Change	2020–2030 % Change	2020 Location Quotient
Columbus, OH	22,621	24,799	2178	10	0.92
Cincinnati, OH, KY IN	20,872	23,584	2712	13	0.84
Cleveland-Elyria OH	15,929	17,277	1348	8	0.67
Dayton-Kettering, OH	13,175	15,362	2187	17	1.51
Akron, OH	5846	6998	1152	20	0.78
Toledo, OH	3949	3969	20	1	0.55
Canton-Massillon OH	1114	1136	22	2	0.29
Youngstown-Warren Boardman, OH-PA	1082	1145	53	6	0.22
Huntington- Ashland, WV-KY-OH	1059	1225	166	16	0.35
Ashland OH	878	1252	374	43	1.84

Source: EMSI, 2020 – QCEW Employees, Non-QCEW Employees, and Self-Employed.

which houses AFRL and a strong concentration of small businesses. The Columbus region has the highest number of R&D jobs, followed by Cincinnati. See Table 4.10.

The workforce in aerospace R&D is dominated by technical professionals. Other than support occupations, most jobs require postsecondary education in a STEM field. The following jobs account for more than 50 percent of the industry, and almost all of them are expected to grow during the next decade. See Table 4.11.

With the growth in the R&D industry across occupations, Ohio will have increasing workforce demands in these STEM fields. Currently, regional completions in STEM fields are keeping pace with the workforce demand. The table below shows R&D occupations across all industries. See Table 4.12.

TABLE 4.11
Top Jobs in the R&D Industry in Ohio

SOC	Description	Employed in Industry Group (2020)	Employed in Industry Group (2030)	Change (2020–2030)	% Change (2020–2030)	Median Hourly Earnings
15-1256	Software Developers and Software Quality Assurance Analyst and Testers	10,549	13,557	3,008	29	$44.26
15-1211	Computer System Analyst	4,124	4,847	723	18	$40.73
17-2051	Civil Engineers	4,034	4,157	123	3	$38.28

(Continued)

TABLE 4.11 ***(Continued)***
Top Jobs in the R&D Industry in Ohio

SOC	Description	Employed in Industry Group (2020)	Employed in Industry Group (2030)	Change (2020–2030)	% Change (2020–2030)	Median Hourly Earnings
15-1232	Computer User Support Specialist	2,988	3,734	746	25	$23.17
17-2141	Mechanical Engineers	2,689	2,791	102	4	$38.24
13-1198	Project Management Specialist and Business Operation Specialist, All Others	2,427	2,744	317	13	$33.77
13-1111	Management Analyst	1,869	2,120	251	13	$39.13
11-3021	Computer and Information System Manager	1,817	2,231	414	23	$61.52
15-1299	Computer Occupation, All Others	1.7 6.4	2,104	340	19	$38.25
41-3091	Sales Representative of Services, Except Advertising, Insurance, Financial Service and Travel	1,756	2,094	337	19	$24.94
11-1021	General and Operations Manager	1,755	2,036	282	16	$46.59
43-9061	Office Clerk, General	1,733	1,745	12	1	$16.49
17-2071	Electrical Engineers	1,691	1,856	166	10	$40.79
19-4021	Biological Technicians	1,661	1,757	95	6	$21.30
13-1161	Market Research Analyst and Marketing Specialist	1,545	1,909	364	24	$29.09
17-2199	Engineers, All Others	1,507	1,567	60	4	$43.73
51-9061	Inspectors, Testers, Sorters, Samplers, and Weighers	1,404	1,192	(212)	(15)	$18.98
11-9041	Architectural and Engineering Manager	1,404	1,479	75	5	$62.51
15-1251	Computer Programmers	1,362	1,407	44	3	$36.88
43-4051	Customer Services Representatives	1,304	1,428	124	10	$16.69

Source: EMSI, 2020 – QCEW Employees, Non-QCEW Employees, and Self-Employed.

TABLE 4.12
Gap Analysis of Top R&D Jobs Across All Industries

Description	2021 Jobs	2030 Jobs	2021–2030 % Change	2021–2030 Change	Avg. Annual Openings	Regional Completions (2019)	Gap
General and Operations Manager	65,608	69,429	6	3,822	5,431	15,156	9,725
Computer and Information Systems Manager	13,933	15,412	9	1,209	1,096	4,448	3,352
Architectural and Engineering Managers	6,214	6,475	4	261	419	9,788	9,369
Management Analyst	28,221	30,574	8	2,352	2,736	10,844	8,108
Market Research Analysts and Marketing Specialists	28,579	31,928	12	3,349	3,068	3,571	503
Project Management Specialists and Business Operations Specialists	47,080	49,421	5	2,341	4,428	171	(4,257)
Accountants and Auditors	51,790	52,968	2	1,178	4,464	3,351	(1,113)
Computer Systems Analysts	27,890	29,211	5	1,321	1,952	2,954	1,002
Computer Network Support Specialists	7,743	8,118	5	376	583	3,156	2,573
Computer User Support Specialists	23,931	25,376	6	1,446	1,828	3,156	1,328
Network and Computer Systems Administrators	12,045	12,506	4	460	775	1,587	812
Computer Programmers	6,014	5,746	(4)	(269)	370	1,896	1,526

(Continued)

TABLE 4.12 *(Continued)*
Gap Analysis of Top R&D Jobs Across All Industries

Description	2021 Jobs	2030 Jobs	2021–2030 % Change	2021–2030 Change	Avg. Annual Openings	Regional Completions (2019)	Gap
Software Developers and Software Quality Assurance Analysts and Testers	47,015	54,261	15	7,246	3,890	3,303	(587)
Computer Occupations, All Other	14,132	14,977	6	845	1,072	2,493	1,421
Aerospace Engineers	3,894	4,057	4	162	244	322	78
Civil Engineers	8,178	8,431	3	253	604	821	217
Electrical Engineers	7,109	7,469	5	360	479	1,349	870
Mechanical Engineers	15,465	16,094	4	628	975	2,251	1,276
Engineers, All Other	8,224	8,478	3	254	532	1,040	508
Architectural and Civil Drafters	3,197	3,157	(1)	(40)	288	230	(58)
Sales Representatives of Services	35,145	36,526	4	1,380	3,991	226	(3,765)
Customer Service Representatives	111,920	112,588	1	668	13,769	1,018	(12,751)
Secretaries and Administrative Assistants	65,941	61,632	(7)	(4,308)	6,541	236	(6,305)
Office Clerks, General	121,365	116,543	(4)	(4,822)	13,068	13	(13,056)
Inspectors, Testers, Sorters, Samplers and Weighers	30,987	27,383	(12)	(3,604)	3,160	7	(3,153)
	791,621	808,489	2	16,868	75,763		

Source: EMSI, 2020 – QCEW Employees, Non-QCEW Employees, and Self-Employed.

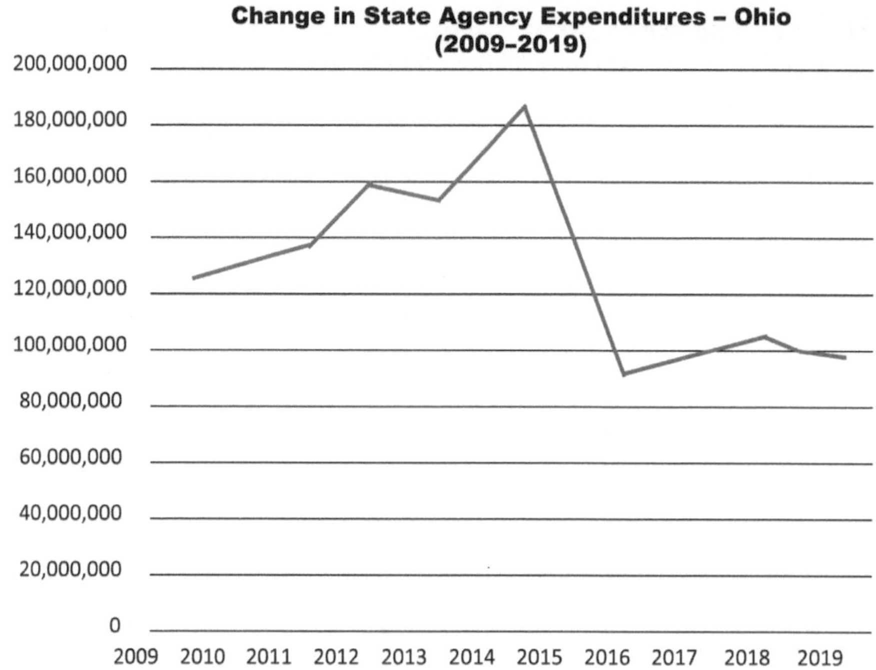

FIGURE 4.7 Change in R&D expenditures by state.

(Source: National Science Foundation, National Center for Science and Engineering Statistics, Survey of State Government Research and Development)

Research Expenditures

Understanding workforce demand in the R&D cluster is only one factor in determining the health and potential growth in the state. Commercialization and entrepreneurship success cannot be measured using traditional metrics like jobs.

Ohio agencies have consistently led the rest of the states in R&D expenditures, and this trend is expected to continue. As some R&D powerhouse states like California have seen a decrease in expenditures, Ohio has continued to increase (see Figure 4.7).

If this trend continues, Ohio has the potential to close the gap with California. As this process occurs, however, strategies must be put into place to leverage innovations resulting from these expenditures. R&D must be converted to commercial application if the state wishes to convert these investments into real economic growth. This could be a bigger challenge for Ohio than any of the other states listed due to the source of these expenditures. Ohio has a disproportional emphasis on companies and individuals contracted to conduct this R&D compared to the top ten states (see Figure 4.8).

In line with R&D expenditures, and with federal organizations like Air Force Research Laboratory and NASA Glenn investing heavily in R&D, an important metric to monitor is Federal obligations. Ohio performs well in this category (in the top 15 states) and could be positioned to improve its federal contract wins (see Figure 4.9).

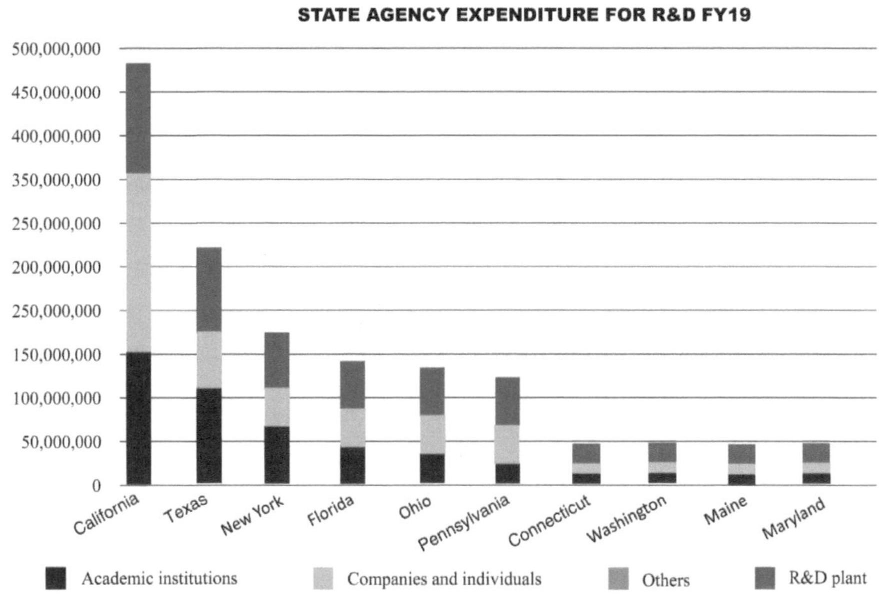

FIGURE 4.8 Top 10 States in R&D expenditures by source.

(Source: National Science Foundation, National Center for Science and Engineering Statistics, Survey of State Government Research and Development. Companies and individuals include individuals under contract for research projects)

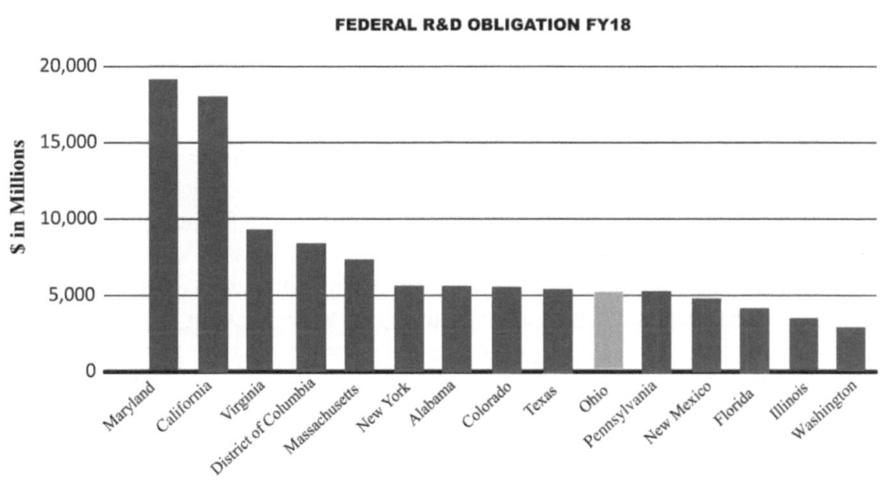

FIGURE 4.9 Federal R&D obligations by state.

(Source: National Science Foundation, National Center for Science and Engineering Statistics, Survey of Federal Funds for Research and Development)

TABLE 4.13
Top States for Federal/Military Jobs

State Name	2021 Jobs	2030 Jobs	2021–2030 Change	2021–2030 % Change	2020 Location Quotient
California	397,137	390,132	(7,004)	(2)	0.76
Texas	352,596	366,937	14,341	4	0.94
Virginia	329,418	328,902	(516)	(0)	2.83
Maryland	217,016	225,308	8,292	4	2.74
Florida	209,423	220,416	10,994	5	0.80
District of Columbia	204,336	196,439	(7,897)	(4)	9.67
North Carolina	186,326	187,547	1,221	1	1.40
Georgia	184,708	189,701	4,993	3	1.39
Washington	142,242	143,903	1,661	1	1.38
New York	131,844	137,521	5,677	4	0.49
Pennsylvania	107,598	105,901	(1,698)	(2)	0.64
Colorado	98,620	98,651	31	0	1.21
Illinois	97,086	95,404	(1,682)	(2)	0.58
Ohio	95,546	97,711	2,165	2	0.62
Hawaii	88,946	91,036	2,090	2	4.46

Source: EMSI, 2020 – QCEW Employees, Non-QCEW Employees, and Self-Employed.

Federal/Military Sector

Ohio ranks 14th in the U.S. for the number of civilian and military jobs. Like many of the other states employing large numbers of civilian and military workers, Ohio's federal jobs are projected to increase slightly. This aligns with federal budgets and expenditures (see Table 4.13).

Understanding occupational demand for federal jobs in Ohio is challenging, since 40 percent of the staffing pattern is made up of military positions which are not disclosed in labor market data. However, of the remaining 60 percent (civilian positions), most of the in-demand positions are increasing (see Table 4.14).

The federal civilian occupations are also in demand in other industries, so a view of occupational demand across all industries is a better view of where workforce shortages may exist. Table 4.15 shows gaps in degree completions related to specific occupations.

Aviation

As the aviation industry is considered a support industry that is driven by consumer demand, it's also an industry that should be monitored. It is an industry that impacts the workforce of other industries, but it is not a driving industry for the state's aerospace and aviation networks (see Table 4.16 and Table 4.17).

TABLE 4.14
Top Jobs in the Federal/Military Industry in Ohio

Description	Employed in Industry Group (2021)	Employed in Industry Group (2030)	Change (2021–2030)	% Change (2021–2030)	% of Total Jobs in Industry Group (2020)
Military-only occupations	18,493	19,456	963	5	19.4
Project Management Specialists and Business Operations Specialists.	5,324	5,372	48	1	5.6
Registered Nurses	2,650	2,682	32	1	2.8
Computer Occupations, All Other	2,355	2,381	6	0	2.5
Personal Service Managers, All Other	1,943	1,985	42	2	2.0
Logisticians	1,798	1,860	63	3	1.9
Management Analyst	1,765	2,006	242	14	1.8
Human Resources Specialist	1,583	1,613	29	2	1.7
Aircraft Mechanics and Service Technicians	1,555	1,615	60	4	1.6
Compliance Officers	1,520	1,540	12	1	1.6
Physicians, All Other and Ophthalmologists, Except Pediatric	1,487	1,490	3	0	1.6
Engineers, All Other	1,302	1,321	19	1	1.4
Office Clerks, General	1,273	1,218	(56)	(4)	1.3
Claims Adjusters, Examiners and Investigators.	1,249	1,165	(84)	(7)	1.3
Buyers and Purchasing Agents	1,153	1,089	(64)	(6)	1.2
Police and Sheriffs Patrol Officers	1,113	1,157	44	4	1.2
Information and Record Clerks, All Other	1,099	1,111	12	1	1.1

(Continued)

TABLE 4.14 ***(Continued)***
Top Jobs in the Federal/Military Industry in Ohio

Description	Employed in Industry Group (2021)	Employed in Industry Group (2030)	Change (2021–2030)	% Change (2021–2030)	% of Total Jobs in Industry Group (2020)
Eligibility Interviewers, Government Programs	1,043	1,039	(4)	(0)	1.1
Lawyers	959	971	12	1	1.0
Transportation Security Screeners	926	934	8	1	1.0
Social Scientist and Related Workers, All Other	900	909	9	1	0.9
Detectives and Criminal Investigators	894	909	15	2	0.9
Social Workers, All Other	882	882	0	0	0.9
Air Traffic Controllers	884	907	22	3	0.9
Airline Pilots, Copilots and Flight Engineers	849	890	41	5	0.9
General and Operations Managers	847	869	22	3	0.9
Licensed Practical and Licensed Vocational Nurses	811	810	(1)	(0)	0.8
Maintenance and Repair Workers General	785	811	26	3	0.8
Accountant and Auditors	721	727	6	1	0.8
Financial and Investment Analyst	682	692	10	1	0.7

Source: EMSI, 2020 QCEW Employees, Non-QCEW Employees, and Self-Employed.

TABLE 4.15
Gap Analysis of Top Ohio Federal/Military Jobs Across All Industries

Description	2021 Jobs	2020 Jobs	2021–2030 % Change	2021–2030 Change	Avg. Annual Openings	Regional Completion (2019)	Gap
General and Operations Managers.	65,608	69,429	6	3,822	5,431	15,156	9,725
Personal Service Managers, All Other.	28,576	30,078	5	1,502	2,082	12,339	10,257
Buyers and Purchasing Agents	18,322	17,576	(4)	(746)	1,629	406	(1,223)
Claims Adjusters, Examiners and Investigators	13,758	13,408	(3)	(350)	1,023	51	(972)
Compliances Officers	8,966	9,462	6	495	750	34	(725)
Human Resources Specialists	28,013	28,924	3	911	2,609	1,162	(1,447)
Logisticians	8,262	8,811	7	550	778	1,284	506
Management Analysts	28,221	30,574	8	2,352	2,736	10,844	8,108
Project Management Specialists and Business Operations Specialists	47,080	49,421	5	2,341	4,428	171	(4,257)
Accountants and Auditors	51,790	52,968	2	1,178	4,464	3,351	(1,113)
Financial and Investment Analysis	16,941	17,805	5	865	1,355	3,342	1,987
Computer Occupations, All Other	14,132	14,977	6	845	1,072	2,493	1,421
Engineer, All Other	8,224	8,478	3	254	532	1,040	508
Social Scientist and Related Workers, All Other	1,803	1,852	3	49	177	216	39
Social Worker, All other	3,443	3,596	4	155	319	2,851	2,532
Lawyers	23,607	23,346	(1)	(262)	1,097	1,352	255
Registered Nurses	129,377	135,923	5	6,546	7,214	15,134	7,920
Physicians, All Other and Ophthalmologists Except Pediatric	18,428	19,381	5	953	631	2,179	1,548
Licensed Practical and Licensed Vocational Nurses	41,610	42,658	3	1,048	3,079	3,248	169
Detectives and Criminal investigators	2,120	2,160	2	40	139	1,428	1,289

(Continued)

TABLE 4.15 *(Continued)*
Gap Analysis of Top Ohio Federal/Military Jobs Across All Industries

Description	2021 Jobs	2020 Jobs	2021–2030 % Change	2021–2030 Change	Avg. Annual Openings	Regional Completion (2019)	Gap
Police and Sheriffs Patrol Officers	25,561	26,120	2	560	1,815	1,358	(457)
Transportation Security Screeners	978	990	1	12	89	0	(89)
Eligibility Interviewers, Government Programs	4,683	4,817	3	133	432	31	(376)
Information and Record Clerks, All Other	3,536	3,675	4	139	389	13	(376)
Office Clerks, General	121,365	116,543	(4)	(4,822)	13,068	13	(13,055)
Aircraft Mechanics and Service Technician	5,424	5,654	4	230	453	258	(195)
Maintenances and Repair Workers, General	61,088	62,595	2	1,507	5,545	38	(5,507)
Airline Pilots, Copilots and Flight Engineers	1,977	1,953	(1)	(24)	197	41	(156)
Air Traffic Controllers	928	952	3	23	79	2	(77)
Military-only occupations	18,493	19,456	5	963	2,048	7	(2,041)
	802,314	823,584	3	21,270	65,671		

Source: EMSI, 2020 – QCEW Employees, Non-QCEW Employees, and Self-Employed.

Unmanned Aircraft Systems

The Unmanned Aircraft Systems (UAS) industry segment is becoming more defined through the advanced air mobility sector. The primary organization used to benchmark data regarding this segment is the Association for Unmanned Vehicle Systems International (AUVSI). AUVSI released an economic impact report in March 2023 titled "The Economic Impact of Unmanned Aircraft Systems Integration in the United States." In this report, researchers applied an aircraft manufacturing staffing pattern to attempt to quantify the impact of UAS integration into the National Air Space (NAS) and used current aerospace activity and infrastructure to project job growth. Using this staffing pattern, they project job growth because of the UAS

TABLE 4.16
Top States for Aviation Jobs

State Name	2020 Jobs	2010 Jobs	2020–2010 Change	2020–2010 % Change	2020 Location Quotient
Texas	89,444	95,854	6,410	7	1.56
California	75,719	85,343	10,624	14	0.93
Florida Virginia	65,630	76,103	10,473	16	1.64
Illinois	42,514	44,031	1,517	4	1.62
New York	39,838	42,931	3,093	8	0.95
Georgia	34,845	35,744	899	3	1.70
North Carolina	21,914	25,891	3,977	18	1.07
Arizona	21,497	22,912	1,415	7	1.63
New Jersey	21,120	23,303	2,183	10	1.18
Colorado	20,149	22,702	2,553	13	1.61
Washington	19,658	23,406	3,748	19	1.25
Michigan	18,475	19,574	1,099	6	0.96
Virginia	16,890	17,157	267	2	0.94
Minnesota	15,512	14,846	(666)	(4)	1.20
Ohio	15,246	15,424	178	1	0.64
Pennsylvania	15,052	15,083	31	0	0.57
Nevada	12,424	15,253	2,829	23	2.00
Massachusetts	11,102	12,527	1,425	13	0.68
Hawaii	10,876	11,853	977	9	3.45
Utah	8,793	10,850	2,057	23	1.28

Source: EMSI, 2020 – QCEW Employees, Non-QCEW Employees, and Self-Employed.

usage for public safety and precision agriculture. One major concern with the use of this staffing pattern is that the UAS that'll likely be produced will be significantly smaller than commercial aircraft and heavily reliant on sensors and other electronic components unique to the systems. This means greater attention on information technology and light-weight polymers would be a good strategy versus expecting existing aircraft engine and parts manufacturers to expand their networks naturally. Ohio performs well using this staffing pattern because of the state's aircraft business; however, it's unlikely the scale and complexity of the engines produced in Ohio would be applicable.

One of the more insightful statements supplied in this report is that "states that create favorable regulatory and business environments for the industry and the technology will likely siphon jobs away from states that do not." While the data supplied in the report places Ohio among the top 15 states poised to grow because of UAS integration into the NAS, these are merely projections based on models, and the state can still rise or fall based on its strategies to capture shares of the manufacturing market. Many of the jobs the UAS will perform may be new jobs that were impossible to accomplish without an unmanned system or these jobs could lead to the

TABLE 4.17
Top Jobs in the Aviation Industry in Ohio

Description	Employed in Industry Group (2021)	Employed in Industry Group (2010)	Change (2021–2010)	% Change (2021–2010)	% of Total Jobs in Industry Group (2020)
Aircraft Mechanics and Service Technicians	2,330	2,404	75	3	15.2
Laborers and Freights Stock and Material Movers, Hand	1,152	1,131	(22)	(2)	7.6
Commercial Pilots	1,041	1,105	64	6	6.8
Airline Pilots, Copilots and Flight Engineers	928	837	(90)	(10)	6.3
Reservation and Transportation Ticket Agent and Travel Clerks	922	855	(66)	(7)	6.2
Flight Attendants	627	582	(44)	(7)	4.2
Customer Service Representatives	529	489	(40)	(8)	3.5
Aircraft Service Attendants and Transportation Workers, All other	346	353	8	2	2.3
Cargo and Freight Agents	335	337	2	1	2.2
First-Line Supervisors of Mechanics, Installers and Repairers	304	310	6	2	2.0

Source: EMSI, 2020 – QCEW Employees, Non-QCEW Employees, and Self-Employed.

loss of jobs that the UAS's purpose would replace. Workforce strategies should be developed to train the workforce already in public safety and precision agriculture to adapt to this disruptive technology (see Figure 4.10).

Building a Workforce Development Ecosystem

This chapter thus far outlines an in-demand industry for a specific region of the country. In addition, this chapter specifically details how a region of the country could analyze an industry and study the demand and supply of its workforce. This

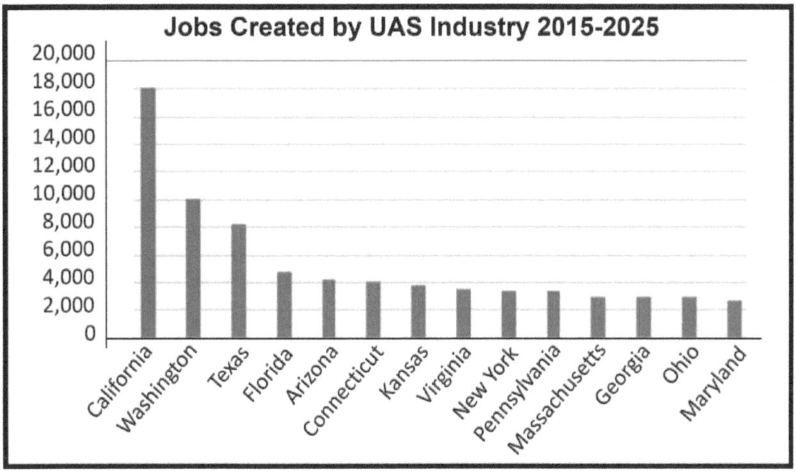

FIGURE 4.10 Projected jobs created by the UAS industry 2015–2025.

(Source: AUVSI, The Economic Impact of Unmanned Aircraft Systems Integration in the United States, March 2023)

type of analysis is the critical first step and ongoing need in building a robust ecosystem to support an industry. An ecosystem in a region includes many different entities. To build a robust ecosystem in support of an industry, it is important to have all stakeholders around the table. For example, the aerospace and defense industry ecosystem would include the following stakeholders:

1. Federal government.
2. State government.
3. Higher Education.
4. Primary & Secondary Education.
5. Industry.
6. Non-profits.
7. Professional Organizations.

The Ecosystem should appoint a strategic leader and develop a strategic focus and vision to be most effective. The goal is to work together in a collaborative way, for the betterment of all in the industry across the region. This initiative is more achievable with collaborative agencies sitting at the table and with incentives layered into the ecosystem through the Federal or State government. A healthy workforce ecosystem can have three states: meeting current industry demand (Steady State), anticipating future industry demand (Predictive State), or strategically driving industry growth (Driver State). Each state has common elements which include investors, producers, consumers, and maintainers; however, the groups that comprise each element can be different for each state.

The lowest energy state is the Steady State focused on meeting current demand, although this still requires healthy collaborations. In the Steady State, investments from government agencies and industry are directed by the source of investment to meet the greatest current gap in workforce supply. Education and local unemployment providers serve as producers to direct the workforce into training programs that will meet the demand. Industry works as consumers, hiring and recruiting talent and ensuring on-the-job training is available to retain the talent. Non-profits serve as maintainers, ensuring the investors and producers are responding to changes in the ecosystem as changes naturally occur. This model ties overall economic development growth to the same growth rate as the existing industry. This can be effective if the existing industry is projected to grow.

The community should always maintain a healthy Steady State, but also develop a strategic vision. This can take two forms: anticipating future industry growth or driving growth. Both states require a robust network of implementers and funding sources with the flexibility to respond to opportunities. Both states represent risk to investors but with the participation of the correct collaborators and the availability of funding with some risk tolerance, these states can have the highest return on investment. The most important consumers are existing businesses, but the source of these consumers is different between the Predictive State and the Driver State.

In a Predictive State, the network is anticipating future growth. A strong economic development arm is needed with the ability to understand national and global industry trends, strong industry partnerships, and the ability to convene both producers and consumers for proactive activity. Critical to this system is the engagement of growing industries and the consumers. They may not be the industries with the strongest representation or the most capital to invest. In this state, the growing industries are not the investors; rather, they are those companies and entrepreneurs connected to the larger network (local, state, national of global networks) of future growth areas. If a critical mass of entrepreneurs can be organized, venture capital becomes the most important currency in this ecosystem, supplemented by the established industries that would benefit the most by the future growth. To ensure government investment, which is an inherently risky adverse, is used with the largest impact, the workforce that would be the most impacted by a shift in industry focus should be identified and training programs should be developed by the providers in anticipation of the entrepreneurs' demand, either through direct jobs or indirect jobs. It's important this workforce investment does not occur too early in the process, training workers for jobs which are not yet available will lead to migration out of the community. The most important consumers in the predictive state are start-ups and the maintainers are those non-profits providing entrepreneurial services.

The Driver State is the ecosystem state with the most energy required, longest timeframe, and highest risk. This model became popular amongst economic development groups employing industry cluster strategies. To be successful, the following is required: flexible capital, information and data-driven strategies, a realistic vision of the time required to reshape the industry sectors, a high tolerance of risk, selective participation of stakeholders, responsive higher education, and industry expertise to aid in strategic planning. Huntsville, Alabama, is a good example of this ecosystem being successful. This region committed to biotechnology growth and STEM education decades before it witnessed the benefits of those efforts. The Huntsville MSA was named the fastest-growing tech hub of 2017 in a study published by ZipRecruiter with a tech job growth rate of 309 percent. In this ecosystem network, public-private partnerships were critical for long-term growth. This long-term growth has taken more than sixty years to realize. The Driver State is a marathon, not a sprint. The strategic plan starts with elementary school programming. In Huntsville, for example, second graders learn to code, and this focus on future skills continues throughout the education system, growing the workforce that will be capable of shaping the targeted industries. Producers in this state become the entire education continuum. Consumers include not just driver industries, but also infrastructure designed to shape the quality of life needed to retain talent. The maintainers are the most important members of this ecosystem. These are regional organizations which have cultivated the networks with the most influence in the community, but also have circles of influence outside the community. This could be the local government, military installations, collaboratives, or strong industry pillars.

For Southwestern Ohio, the aerospace and defense industry may be an attractive target for the Driver State model. Like Huntsville, the federal installation R&D has the potential to produce the technology of the next century, but it must be harnessed soon. Many attempts have been made to tap into the intellectual capital in AFRL represented by thousands of unlicensed patents. Continued efforts to break into this space are needed, and perhaps the best option for the region is to focus on a Predictive State, growing the entrepreneur space, in anticipation of a successful Driver State strategy.

To do this, leveraging the growth in the R&D sector will require a shift in business practices and using network models that enable exponential growth. For a network model to be successful, all nodes within the network must both give and receive value. From the organizations conducting research all the way to the customer who will buy a commercial product, all participants in the commercialization process must create value.

With respect to the potential loss of STEM graduates produced in Ohio, but not hired in Ohio, the best strategy to retain those workers is to connect them to nodes in a network for them that will be valuable before they graduate. In a recent report, a large IT company reported 80 percent of its U.S. workforce was white. This stark data point speaks more about networks than it does to inclusion. Companies will hire from programs they're familiar with and students they know are up for the task. As a summarizing view, Figure 4.11 illustrates the elements of Steady State, Predictive State, and Driver State that could serve as anchors for innovation and local business development.

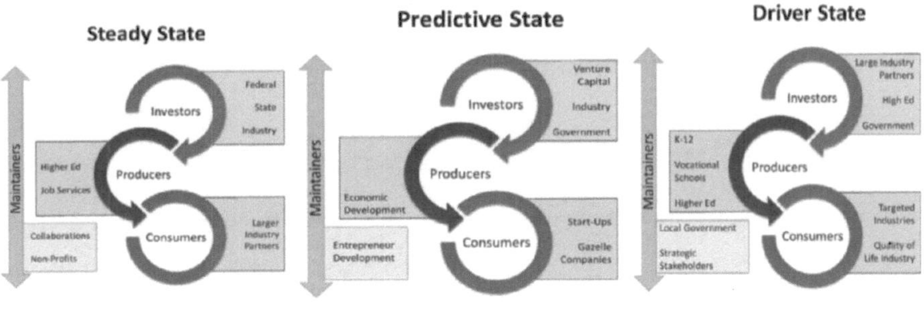

FIGURE 4.11 Graphics of steady state, predictive state, and driver state.

REFERENCES

AIA (2021), https://www.aia-aerospace.org/publications/2021-facts-figures/

Aerospace Industries Association (December 2023), "The State of the US Aerospace and Defense Industry."

Association of Unmanned Vehicle Systems International (2023), "The Economic Impact of Unmanned Aircraft Systems Integration in the United States."

Barlow, C. B., and K. Rochon (2020), "The Aerospace and Defense Industry in Southwest Ohio: A Model for Workforce-Driven Economic Development," in **Defense Innovation Handbook: Guidelines, Strategies, and Techniques**, CRC Press/Taylor and Francis, Boca Raton, FL, pp. 39–64.

Boeing (2023), "Aerospace Services Market Outlook." https://www.boeing.com/commercial/market/commercial-market-outlook#overview

Boeing (2024), https://www.boeing.com/commercial/market/commercial-market-outlook#overview

Bureau of Labor Statistics databases. https://www.bls.gov/data/. Accessed July 30, 2025.

National Science Foundation Science and Engineering Indicators State Data Tool. https://ncses.nsf.gov/indicators/data

Schaeffer, Leonard D. (2024), https://schaefferfellows.org/annual-report/2024-summit/, Accessed July 28, 2025.

U.S. Census Bureau, American Community Survey. https://www.census.gov/programs-surveys/acs.html

Ziprecruiter (2023), "Fastest Growing Tech Towns in 2023." https://www.ziprecruiter.com/blog/ten-up-and-coming-tech-towns/

5 Work Design for Manufacturing Workforce*

BACKGROUND

The need for the next generation of manufacturing design as discussed in Bommer (2017) is summarized below:

> The demands of manufacturing processes and design are evolving rapidly. Production facilities must reconfigure processes and adapt swiftly to changing market demands to remain competitive and profitable in a global market. Consequently, production employees face increased responsibilities, but more complex procedures and automation can heighten their cognitive load. Traditional manufacturing process design primarily focuses on physical ergonomics, aiming to reduce physical fatigue and discomfort to enhance throughput and minimize safety risks. However, this approach may not suffice for modern manufacturing needs. Effective system ergonomics now requires incorporating cognitive evaluations into process design. As a result, the next generation of manufacturing systems should address both physical and cognitive ergonomics.

This chapter covers mental workload measures for work evaluation to assess levels of mental load imposed by a system. Mental workload (MWL) is an area of study within cognitive ergonomics. "Cognitive Ergonomics is the scientific discipline that studies the cognitive processes in the design of technology and the environment in which this technology is used by people." (Cañas, 2008). MWL measures are major factors in the system design and analysis of occupational tasks (DiDomenico & Nussbaum, 2011). MWL assessments can be used to estimate performance requirements for the design of systems and to predict workload changes with various system modifications (May et al., 1990). Many modern systems are made up of increased human demands by requiring individuals to perform physically demanding tasks simultaneously with more cognitive responsibilities. As these procedures for human-machine systems are becoming more complex and automated, the need for mental workload assessment is becoming more critical (Rubio, Díaz, Martín, Peunte, 2004). To develop improved work systems that are globally competitive, the process designer should consider MWL evaluations for the development of optimal work performance and error reduction.

* *Reprinted and Adapted from Bommer and Sharon (2017)*

DOI: 10.1201/9781003537755-5

INFORMATION PROCESSING

Information processing is extremely important to human performance because in situations where humans interact with work systems, "the operator must perceive information, must transform into different forms, must take actions on the basis of the perceived and transformed information, and must process the feedback from that action, assessing its effect on the environment." (Wickens & Carswell, 2006). This information is received visually or via an auditory organ (perception), and must be understood in order to interpret the meaning of the perceived information so decisions can be made interacting with the knowledge stored in the memory system (cognition) (Murata, 2000).

According to the Wickens and Holland (2000) framework for information-processing, sensory memory, working memory, and long-term memory are the three stages of information processing. Sensory memory is associated with interpreting data based up past experiences that were sensed or perceived. However, working memory (i.e., short-term memory) is a brain system that deals with immediate conscious processing of data; it is of limited capacity and heavily demanding of attention to information that was not sensed or perceived but was generally an internal thought. An example of working memory utilization is quality inspection tasks (recalling memorized criteria for acceptable quality) that impose sustained load on the working memory. This demand may fatigue the operator or otherwise affect performance in the quality inspection task (Pesante et al, 2001). Long-term memory is the cognitive storage of data; it keeps a vast amount of information but it's not always able to be totally recalled. It is important that the operator is not over-saturated with tasks that can cause a mental overload situation because overloading the operator could lead to performance degradation. Mental workload or simply 'workload' is the interaction between an operator and his or her assigned task in a human-machine system. In simple terms, it is the demand on an operator.

MENTAL WORKLOAD (MWL)

Mental workload (MWL) is defined as a multidimensional model that encompasses task and performance demands along with operator skill and attention (Stanton et al., 2004). MWL measures offer awareness as to where increased task demands could lead to an adverse influence on human performance. The demands of a task may require completing physical activities and/or performing cognitive tasks. Understanding the task demands is important for work design in order to design processes to reduce stress levels, accident rates, and errors. When the sum or difficulty of tasks required to complete a process increases, or when time allocated for task completion decreases, workload generally rises (Colombi et al., 2012). Therefore, increased safety, health, comfort, and long-range productivity can be accomplished by regulating task demands, such that the operator is neither underloaded nor overloaded (Rubio, Díaz, Martín, & Puente, 2004). This is relevant to work design because high cognitive load for extended periods of time can lead to unproductive processes and poor performance as well as ergonomic and mental health symptoms (Lindblom & Thorvald, 2014).

Mental workload can affect an operator's ability to support or attain desired performance levels (Xie & Salvendy, 2000). The concept of mental workload deals with the difference between the amount of resources available to a person and the quantity of resources necessary for the task. (Jung & Jung, 2001) For this reason, high workload levels occur when the task demands exceed the operator's mental resource capacity (Loft et al., 2007). Therefore, mental workload can be modified by altering the amount of resources available within the person or the demands required by the task on the operator. The Multiple Resource Theory (MRT) is a model that can be employed to assess the various mental resources for various conditions in the design of the human-machine system.

According to Estes (2015), workload measurements became an area of significant research in the late 1960s, when a variation of techniques were being produced to measure it. Workload measurements have been utilized in variant functions to include system desirability and system optimization. As the operator works the system, either with a simulation or the existing process, the operator's mental workload can be evaluated and measured. MWL can be fragmented into two segments: effective and ineffective. Effective workload (i.e., task load) is the minimal amount of workload produced by task requirements. The operator directly accomplishes the work with this portion of the workload (Xie & Salvendy, 2000). In contrast, ineffective workload can be avoided by workers because it does not directly contribute to completing a task (Xie & Salvendy, 2000). Ineffective workload can be lowered by learning and training. As operators' skill levels increase, their proficiency at completing a task increases and their perception of workload decreases (Colombi et al., 2012). When executing tasks, effective workload corresponds to fast and accurate actions, whereas ineffective workload is associated with error and inaccuracy. Effective workload necessitates the smallest amount of attention because it correlates to efficiency; however, "more attention is needed under the ineffective mode to control each stage of information processing." (Xie & Salvendy, 2000) As shown below in Figure 5.1, there are three categories of MWL measures: subjective, physiological and performance. These different measures assess different dimensions of

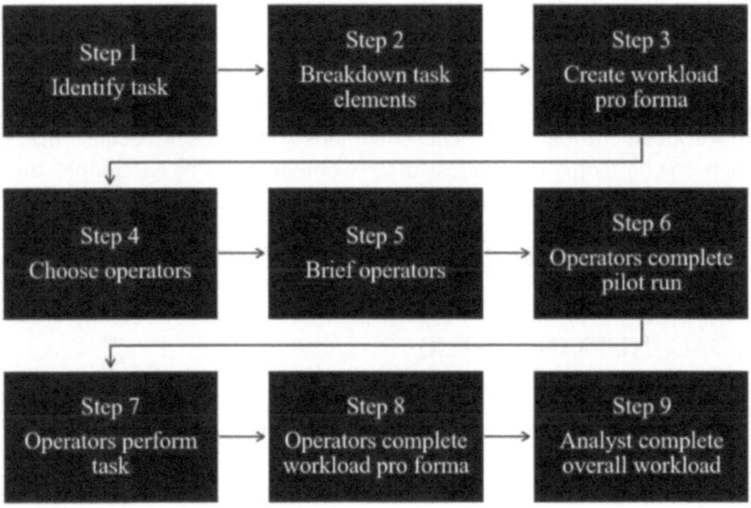

FIGURE 5.1 Measures for mental workload analysis.

mental workload, so it is advisable to use multiple measures for better diagnosticity (Miller, 2001; Tsang & Vidulich, 2006).

SUBJECTIVE MEASURES

Subjective workload measures involve the operator by providing his or her judgment on the cognitive effort entailed to accomplish a task after completing it (Cegarra & Chevalier, 2008; Stanton et al, 2010; Hertzmun & Holmegaard, 2013). These assessments are based upon an individual's perception of the effort applied in performing a task. This method is typically not complicated to administer and has gained considerable theoretical support for its ability to provide sensitive measures on operator workload (Rubio, Díaz, Martín, & Puenete, 2004). As cited by Bommer and Fendley (2015), subjective mental workload measures have been widely accepted and these measurement techniques have proven to be valid and reliable.

Subjective workload measures attempt to quantify the effort utilized while performing a task. They use numerical ratings that will not measure task performance or physiological responses associated directly with the work applied (DiDomenico & Nussbaum, 2008). This method, in general, is a good indicator of overall workload assessment because most of these techniques will not pinpoint or differentiate differences in loading peaks. There are a host of subjective methods to measure MWL that provide useful information. These include:

- Cooper-Harper Scales (Cooper & Harper Jr, 1969)
- The Bedford Scale (Roscoe & Ellis, 1990)
- Subjective Workload Dominance Technique (Vidulich, Ward & Schueren, 1991)
- Subjective Workload Assessment Technique (Reid & Nygren, 1988)
- NASA – Task Load Index (Hart & Staveland, 1988)
- Workload Profile (Tsang & Velazquez, 1996).

In this chapter, two subjective mental workload methods will be discussed: National Aeronautics and Space Administration – Task Load Index (NASA-TLX) and the Workload Profile (WP). These methods were chosen because the NASA-TLX has been commonly used across diverse domains. WP is based upon the sound underpinning theory of Multiple Resource Theory (MRT), which can aid in pinpointing the mental resource causing the workload.

NATIONAL AERONAUTICS AND SPACE ADMINISTRATION – TASK LOAD INDEX (NASA-TLX)

NASA-TLX is an MWL assessment approach that is the most applied to the subjective category. It has been utilized in various domains to include civil and military aviation, driving, nuclear power plant control room operation, and air traffic control (Stanton et al, 2010) and has been cited in previous studies, including Endsley & Rodgers (1997), Bertram et al (1992) and Cao et al (2009) Its application has expanded beyond its original application of aviation and crew complement. It is "being used as a benchmark against which the efficacies of other measures, theories, or models are judged" (Hart, 2006). It has been taught in university courses and used in analyzing diverse environments, such as aircraft certifications, operator rooms, and nuclear power plant control rooms.

The NASA-TLX uses a multi-dimensional scale to quantify task performance of the operator. Its multi-dimensional scale consists of six subscales:

- Mental demand.
- Physical demand.
- Temporal demand.
- Effort.
- Performance.
- Frustration level.

The mental demand measures the perceptual elements of the task to include thinking, looking, and searching activities. The physical demand evaluates whether the physical activity was laborious or relaxed. The temporal demand measures the presence of how much pressure is perceived by the pace of the activity. Effort assesses how challenging the work was to accomplish the tasks. The performance dimension considers how content the operator was with his or her execution of accomplishing the task. Lastly, the frustration level judges how the operator felt while completing the task.

Figure 5.2 outlines the seven steps to implementing the NASA-TLX. Step one of the process is to identify the task subject to analysis. The next step requires breaking down the work elements of the task by conducting a Hierarchical Task Analysis (HTA). During step three, operators are chosen to meet the needs of the experimental design or process modification to be analyzed. Once the operators have been selected, they should be briefed (step four) on the aim of the evaluation and the

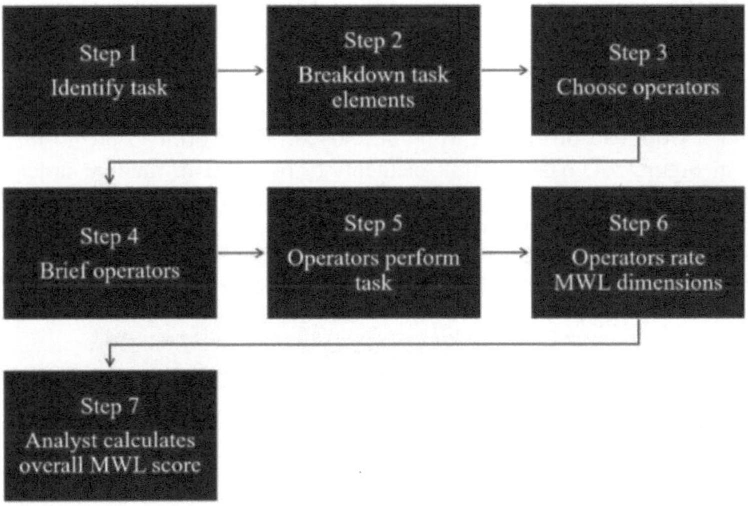

FIGURE 5.2 NASA-TLX procedural steps.

NASA-TLX method. In step five, the task under analysis is being performed by the operator. After this task is completed, the operator starts the NASA-TLX rating procedure. It is recommended that the operators be given the definitions of each MWL dimension with the rating scale to assist with this process. This step includes completing a pair-wise comparison and interval scale. The pair-wise comparison is a weighting procedure that poses fifteen pairwise combinations to the operators and has them choose the scale from each pair that has the most influence on their perceived workload during the task under analysis (Stanton et al., 2010). The most common modification to the NASA-TLX has been to exclude the weighting process from the procedure (Hart, 2006). The modified method has been referred to as the Raw NASA-TLX (RTLX); the sub-scale ratings are averaged to form an overall workload appraisal. The interval scale rates the six dimensions on a scale from 0 to 100, where 0 represents the lowest task demand and 100 is highest task demand except for the performance dimension, where 0 indicates high demand and 100 is low demand. The final action of the process (step seven) is to calculate the overall workload score.

Workload Profile

The Workload Profile (WP) is a multidimensional subjective workload measurement tool that is based on the multiple resource model proposed by Wickens (1987), Tsang and Velazquez (1996), Stanton et al. (2010), and Rubio et al. (2004). This assessment tool utilizes eight dimensions in the collection of data on the demands imposed by a task: perceptual/central processing, response selection execution, spatial processing, verbal processing, visual processing, auditory processing, manual output, and speech output. The WP technique offers an overall rating for the specific task elements as well as a rating for each of the workload dimensions. These multidimensional ratings should provide the analyst with a helpful interpretation of MWL to categorize the means in which the task is demanding (Tsang & Velazquez, 1996).

Figure 5.3, previously shown, summarizes the nine steps of the Workload Profile Process as cited by Stanton et. al (2010). The first step of the process is to specify the tasks or scenarios under analysis. Next, an HTA should be administered to break down the work elements. This aids in understanding the complete task. Step three calls for a workload pro-forma to be developed. The workload pro-forma, shown below in Figure 5.4, lists the task elements to be rated in random order down a column and records the eight MRT workload dimensions across the page. These dimensions are stages of processing (perceptual/central vs. response), modalities of

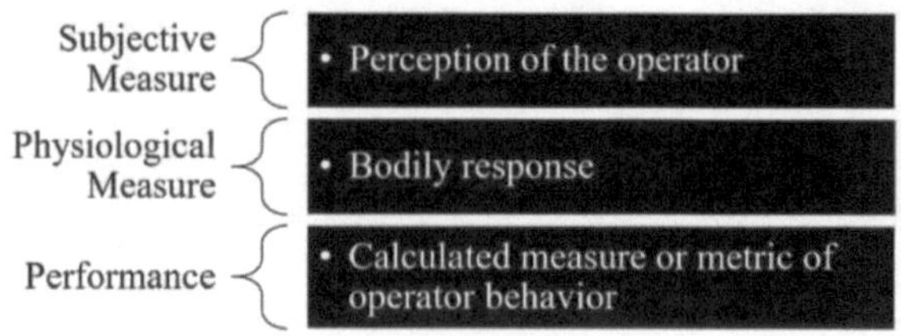

FIGURE 5.3 Workload profile procedural steps.

Operator #		Workload Dimensions							
		Stages of Processing		Code of Processing		Input		Output	
Work Elements		Perceptual	Response	Spatial	Verbal	Visual	Auditory	Manual	Speech
#1									
#2									
#3									
#4									

FIGURE 5.4 Workload profile pro-forma (adapted from Tsang & Velazquez, 1996).

input (visual vs. auditory), output (manual vs. vocal), and codes of processing (verbal vs. spatial).

The operators are chosen in step four. In step five, the operators are briefed on the purpose of the process evaluation. It is recommended that they receive a seminar or lesson on MWL, MWL assessment, MRT, and the workload profile dimensions, to gain a clear understanding of the process. One disadvantage of the WP method is that understanding the different dimensions is difficult, if there is limited experience in psychology and human factors. Once the operators have a clear understanding of the technique's process, in step six, a pilot run with a small set of work elements should be completed to demonstrate how to complete the profile pro-forma. The task under analysis will be performed by the operators in step seven. Step eight requires the completion of the workload profile pro-forma. During this step, the operators rate the proportion of attentional resources used for each task. Ratings ranging from zero to one for each of the MWL dimensions hypothesized in Wicken's model of the Multiple Resource Theory are assigned by the operators. During the final step (step nine), the analyst calculates the overall workload for each task by adding the ratings across all of the dimensions.

Multiple Resource Theory (MRT)

The underpinning theory of WP is the Multiple Resource Theory (MRT). MRT is a predictive model that provides an understanding of operator performance abilities while multi-tasking in a complex environment (Wickens, 2002). Complex environment describes a task that has several concurrent work elements that are time-shared (Liu & Wickens, 1988). When an individual performs a task, each operation deploys mental processing resources vital to carrying out the task (Mitchell, 2003). People have the capacity to multi-task until task demands exceed their available resources. According to MRT, the human mind has various resources that can be allocated to task demands either individually or collectively to include visual, auditory, cognitive, motor, and speech. When task demands overlap, available resources are reduced and MRT model predicts that performance will degrade. When multiple tasks involve competing resources, this competition could lead to a compromise in system safety and effectiveness. So, the principal value of the MRT model is predicting relative differences in multitasking amongst various conditions (Wickens, 2008).

The MRT model can predict performance breakdowns in high workload circumstances, such as tasks that require an operator to multitask by performing two or more work elements at one time. This model can make practical predictions regarding performance breakdowns of a human operator's ability to accomplish tasks like driving an automobile in traffic, a pilot landing an aircraft, or a secretary in a busy office (Wickens, 2002). A significant application of MRT is its ability to identify resource overload when conditions of multitasking occur. Examples of strategies for reducing resource overload include automating parts of tasks, reassigning task elements among multiple operators, and changing task procedures, so process steps are performed sequentially instead of concurrently (Wickens et al, 2015).

The MRT model uses four dimensions that justify the variance in time-sharing of performance and each dimension has two discrete levels (Wickens, 2002). Wickens (2008) defines the dimensions as:

- Processing stages.
- Perceptual modalities.
- Visual channels.
- Processing codes.

During the processing stages, perceptual and cognitive tasks utilize different resources from the selection and execution of tasks. Individuals are capable of time-sharing effectively between activities that require perceptual and cognitive demand because these resources appear to be the same. The perceptual modalities dimension infers that auditory perception applies more resources than visual perception. The implication of this dimension is that time-sharing for cross-modal conditions versus intra-modal conditions is better. This is because attention can be distributed better between the eye and ear rather than two visual or two auditory channels, since visual perception uses different resources than auditory perception. Visual channels differentiate between two different types of resources: focal and ambient vision. Focal vision supports object recognition, and ambient vision is responsible for the perception of orientation and movement of objects. Processing codes indicate that spatial activity employs different resources than verbal/linguistic activity. Separating the spatial and verbal resources improves efficiency. Understanding the dimensions in the MRT model can aid in evaluating and allocating the various mental resources for different task conditions in the work system design. There are simulation tools, such as the Improved Performance Research Integration Tool (IMPRINT) and the Man-Machine Integration Design and Analysis System (MIDAS), that are based upon MRT and can be used for workload modeling (Wickens, 2008).

PHYSIOLOGICAL MEASURES

Physiological workload measures are applied mostly to uninterrupted measurements of the body's physical responses (Miller, 2001). An advantage of these measures is that they do not interfere with the primary-task performance, but depending upon the collection method, there may be a degree of physical constraint imposed (Wickens et al, 2015). The use of physiological measures as an indicator of MWL is based upon the assumption that as task demand increases, changes in various participant physical responses are noticeable (Stanton et al., 2010). These physical responses can be measured with various bodily activity:

- Cardiac.
- Brain.
- Respiratory.
- Speech.
- Eye.

Cardiac activity is generally measured with the operator's heart rate, heart rate variability (HRV), and blood pressure. Brain activity can be measured using an electroencephalograph (EEG) and electrooculogram (EOG). To measure respiratory activity, the volume of air a person breathes in and the number of breaths taken in a given time can be used. Speech measures of pitch, rate, and loudness can be taken into account when evaluating mental workload. Eye measures can include fixation rate, fixation duration, blink rate, blink duration, pupil diameter, saccade duration, and saccade frequency.

As previously mentioned, the use of physiological measures can be evaluated by various methods; however, this review will focus on eye activity. For over twenty years, psychologists have argued that pupillary changes are indications of cognitive load, and many studies have validated this argument (Marshall, 2002). Fixations are moments when the eyes are looking at an object. Blink rate is the number of times the eyes close in a specified time. Blink duration is an interval of eye closure defined as the length of time spent blinking. Pupil diameter has been found to increase with an increasing workload, so as the size of the pupil increases with a higher workload. Saccade is the involuntary eye movement between fixations. According to Ahlstrom and Friedman-Berg (2006), research findings indicate that blink rate and blink duration decline as workload decreases, pupil dilation increases as cognitive demand increases, and an increase in the number of saccades or decreased saccade duration was found relative to increased workload. Also, increased fixation duration can be interpreted as an indicator of increased mental workload. Doherty et al (2010) found that the frequency of long fixations increases with increased workload.

To collect the eye measurements, the process of eye-tracking is utilized. Eye tracking is a technology used for measuring eye activity, such as measuring where an individual is looking (i.e., gaze direction) at any given time, the sequence in which the eyes are shifting from one location to another, or how much the pupil dilates while looking. An eye tracker is a device that deals with the measuring of eye positions and eye movement. Eye trackers are used to evaluate human-machine interaction. The eye-tracking data is collected using either a remote, head-mounted, or wearable eyeglass system. This data can be utilized to collect information relative to numerous gaze measurements. Since eye movements are commonly considered to be involuntary, eye tracking provides objective data of users' visual interaction with a system (Bruneau et al. 2002).

PERFORMANCE

Mental workload can be measured by monitoring an operator's performance and observing how it varies with changes in task demands (Hertzum & Holmegaard, 2013). Estimations of mental workload can be determined by measuring a person's performance on a task and evaluating whether his or her performance degrades with increasing workload (Miller, 2001). When considering performance for repetitive task processes, such as manufacturing operations, not only should the calculated measures of the operator's performance be utilized to evaluate MWL, but the factors of unconscious cognition and learning curves should also be taken into account for the system design.

CALCULATED MEASURES

The key categories of performance-based workload measures are primary-task performance and secondary-task performance. The primary-task performance measures the operator's adequacy of performance in the main task, while the secondary-task performance evaluates the workload of the primary task of interest by assessing the operator's ability to perform a second task in conjunction with the primary task. According to Miller (2001), the primary-task performance should be used to measure mental workload because the primary-task performance is easier to measure than secondary-task performance, and it has been extensively studied for accuracy. Also, it is expected that the primary task performance of the operator will reduce with increased MWL for the task under analysis (Stanton et al, 2010).

The mental workload necessitated for completion of a given task can be assessed with performance metrics, such as time taken, speed, or number of errors (Cegarra & Chevalier, 2008). As cited by Cegarra and Chevalier (2008), studies have shown a correlation between poor performance in the main task and the mental workload experienced by participants. Bubb (2005) defined human error probability (HEP) as the likelihood that a task would be completely defective during a specified timeframe. HEP is an accuracy measurement expressed mathematically as follows:

$$\text{HEP} = \frac{\text{number of observed errors}}{\text{number of possiblities for an error}}$$

Galy, Cariou, and Mélan (2011) cite a study by Fournier et al (1999), which examined subjects in a multi-tasking experiment. The results revealed that performance decreased as workload increased. This implies that a decline in performance can be an indicator of MWL overload. Therefore, the HEP metric should increase as MWL increases. The HEP metric is only one example of a performance measure. Others can be defined depending on the goal of the process design.

UNCONSCIOUS COGNITION

In repetitive task operations, like production processes, there is the process of learning. Learning is a natural phenomenon, in which a worker's output improves as his productivity levels improve (Badiru, 1995). As an operator becomes more efficient in a process over time and the learning reduces, an operator can experience unconscious cognition. Unconscious cognition is the "mental structure and process that, operating outside phenomenal awareness, nevertheless influences conscious experiences, thought and action" (Kihlstrom, 1987). An example of unconscious cognition is the actions required while driving an automobile. These actions, which are completed regularly without apparent mental effort and regularly without the ability to remember what was done, are considered by some as being done unconsciously (Greenwald, 1992). If one does experience unconscious cognition, according to Reinhard et al (2013), unconscious processes are presumed to have far better processing capability than conscious thinking. As a result of this unconscious state, one should outperform conscious thought when decision-making problems are complex. Therefore, the ability to design a work system to achieve unconscious cognition levels for the operator would be the most desirable system design.

LEARNING

"Learning refers to the improved operational efficiency and cost reduction obtained from repetition of tasks." (Badiru, 2010, 2012) Workers learn and improve work performance by repeating tasks over time. As workers become more skilled and develop higher capabilities, a larger mental resource supply is effectively created, improving the efficiency of the operator (Vidulich & Tsang, 2012). Therefore, their output levels should improve with learning, which decreases expenditures per unit and believably improves product quality (Badiru, 1995). However, a worker's performance can also regress as a result of forgetting and work interruptions. Forgetting is an outcome of work interruptions, which becomes apparent by production rate reductions after an inoperative period and the manufacturability of lower quality products compared to production during continuous operation (Jaber, 2011). Therefore, the ability to predict an operator's performance when work resumes is valuable because this can improve production planning and resource allocation. In addition, incorporating learning into a production process decreases fatigue, which can also improve the performance of the system (Jaber et al, 2013).

"The learning curve phenomenon has been of interest to researchers and practitioners for many years." (Jaber, 2011) Learning curves are used to determine how fast a skill can be mastered. However, industrial learning curves can be used for production planning and control, and there is a large body of literature examining how processes have been improved as a result of learning curves. Learning curves have been used in manufacturing for decision-making, manpower scheduling, resource allocation, cost optimization, and more. Although learning is a well-examined topic, this concept should be further extended to study and experiment to understand how forgetting affects mental workload. From literature, it is known that after a break the first cycle of work is the longest; and, according to mental workload theory, performance (e.g., time on task) is an indication of cognitive load. This correlates to an operator experiencing the greatest load at the first cycle of work. Using learning curves to determine learning profiles could assist in reducing an operator's cognitive load at the start of production.

To model a worker's output, learning curves can be utilized. Learning curve analysis can be useful for predicting expected work rates in processes where output accountability is tracked (Badiru, 2014). In jobs with repetitive tasks, learning curves (LC) can mathematically depict a worker's performance (Jaber, 2011). According to Jaber (2011), this method is deemed an effective tool to monitor workers' performance in repetitive tasks, leading to a reduction in process loss due to the workers' inability during the first production cycle. LCs can be applied for the allocation of tasks to workers consistent with their learning profiles to control productive operations. Other applications of LCs include production planning, cost estimation and control, resource allocation, lot sizing, and product pricing (Badiru, 1995). High performance with low mental effort is the most efficient combination of cognitive capacity, which can result from the theory of learning.

SUMMARY

Processes should be evaluated to confirm whether high levels of mental workload are within worker's capabilities, so that the average workload does not go so high that it creates cumulative mental fatigue and not go so low that operators are bored and may miss important system cues (Resnick, 2014). There are various mental workload measures that can be used to confirm workload levels imposed by the system. These mental workload measures are often used to verify whether the human is operating within a tolerable information processing capacity while executing task functions and elements. As technology continues to change, there is a constant need to know its impact on the human operator to design safe and highly productive operations for work performance. Workload is the quantity of resources applied in the performance of a specific task (Boff & Lincoln, 1988). The mental resources of humans are limited; therefore, it is important that the work performed does not overload human capacity, which could lead to performance degradation. To optimize mental workload in efforts to improve performance, the analyst must have accurate predictions of the performance. There are three categories of MWL measures: subjective, physiological, and performance. No single measure of MWL evaluates all the dimensions of mental load; so, when possible, use a combination of the measures from the three major categories. However, for maximum accuracy of mental workload measurement, it is recommended that one measure be continuous, one measure be taken at intervals during the experiment, and one measure be taken after the experiment (Miller, 2001). Once an accurate mental workload prediction is obtained for a task, these predictions can be used for overall capacity planning of work systems by identifying the need for mental workload reductions, as well as determining if additional work elements can be assigned to operators without degrading performance. The multiple resource theory can be employed to determine which mental resources were used in efforts to enhance the process design by restructuring the mental resource utilization to improve operator performance. See Vidullch et al, 1991; Jaber and Bonney, 2011 for further and related studies.

REFERENCES

Ahlstrom, U., and F. J. Friedman-Berg (2006), "Using Eye Movement Activity as a Correlate of Cognitive Workload," ***International Journal of Industrial Ergonomics***, Vol. 36, No. 7, pp. 623–636.

Badiru, A. (2014), *Adverse Impacts of Furlough Programs on Employee Work Rate and Organizational Productivity*,

Badiru, A. B. (1995), "Multivariate Analysis of the Effect of Learning and Forgetting on Product Quality," ***The International Journal of Production Research***, Vol. 33, No. 3, pp. 777–794.

Badiru, A. B. (2010), "Half-life of Learning Curves for Information Technology Project Management," International Journal of IT Project Management, Vol. 1, No. 3, pp. 28–45.

Badiru, A. B. (2012), "Half-Life Learning Curves in the Defense Acquisition Lifecycle," Defense Acquisition Research Journal, Vol. 19, No. 3, pp. 283–308.

Badiru, A. B. and S. C. Bommer (2017), Work Design: A Systematic Approach, Pp. 121-136, CRC Press/Taylor and Francis, Boca Raton, FL.

Bertram, D. A., D. A. Opila, J. L. Brown, S. J. Gallagher, R. W. Schifeling, I. S. Snow, et al. (1992), "Measuring Physician Mental Workload: Reliability and Validity Assessment of a Brief Instrument," ***Medical Care***, Vol. 30, No. 2, pp. 95–104.

Boff, K. R., and J. E. Lincoln, editors (1988), Engineering Data Compendium: Human Perception and Performance. Laboratory User's Guide, University of Michigan, Ann Arbor, MI

Bommer, S. C. (2017), Cognitive Ergonomics in Manufacturing. **ISE Magazine**, Vol. 49, No. 1, pp 41–43.

Bommer, S. C. and M. Fendley (2015), "Assessing the Effects of Multimodal Communications on Mental Workload During the Supervision of Multiple Unmanned Aerial Vehicles," ***International Journal of Unmanned Systems Engineering***, Vol. 3, No. 1, p. 38.

Bruneau, D., M. A. Sasse, and J. McCarthy (2002). "The Eyes Never Lie: The Use of Eye Tracking Data in HCI Research," **Proceedings of the CHI**, Vol. 2, pp. 25.

Bubb, H. (2005), "Human Reliability: A Key to Improved Quality in Manufacturing," ***Human Factors and Ergonomics in Manufacturing & Service Industries***, Vol. 15, No. 4, pp. 353–368.

Cañas, J. (2008), "Cognitive Ergonomics in Interface Development Evaluation," ***Journal of Universal Computer Science***, Vol. 14, No. 16, pp. 2630–2649.

Cao, A., K. K. Chintamani, A. K. Pandya, and R. D. Ellis (2009), "NASA TLX: Software for Assessing Subjective Mental Workload," ***Behavior Research Methods***, Vol. 41, No. 1, pp. 113–117.

Cegarra, J., and A. Chevalier (2008), "The Use of Tholos Software for Combining Measures of Mental Workload: Toward Theoretical and Methodological Improvements," ***Behavior Research Methods***, Vol. 40, No. 4, pp. 988–1000.

Colombi, J. M., M. E. Miller, M. Schneider, M. J. McGrogan, C. D. S. Long, and J. Plaga (2012), "Predictive Mental Workload Modeling for Semiautonomous System Design: Implications for Systems of Systems," ***Systems Engineering***, Vol. 15, No. 4, pp. 448–460.

Cooper, G. E., and R. P. Harper Jr (1969). *The use of Pilot Rating in the Evaluation of Aircraft Handling Qualities*, Research Report 567, Ames Research Center, Moffett Field, CA

DiDomenico, A., and M. A. Nussbaum (2008), "Interactive Effects of Physical and Mental Workload on Subjective Workload Assessment," ***International Journal of Industrial Ergonomics***, Vol. 38, No. 11, pp. 977–983.

DiDomenico, A., and M. A. Nussbaum (2011), "Effects of Different Physical Workload Parameters on Mental Workload and Performance," ***International Journal of Industrial Ergonomics***, Vol. 41, No. 3, pp. 255–260.

Doherty, S., S. O'Brien, and M. Carl (2010), "Eye Tracking as an MT Evaluation Technique," ***Machine Translation***, Vol. 24, No. 1, pp. 1–13.

Endsley, M. R., and M. D. Rodgers (1997). *Distribution of Attention, Situation Awareness, and Workload in a Passive Air Traffic Control Task: Implications for Operational Errors and Automation*, Research Report DOT/FAA/AM 97/13, US Department of Transportation, Washington, DC.

Estes, S. (2015), "The Workload Curve: Subjective Mental Workload," ***Human Factors***, Vol. 57, No. 7, pp. 1174–1187.

Fournier, L. R., G. F. Wilson, and C. R. Swain (1999), "Electrophysiological, Behavioral, and Subjective Indexes of Workload When Performing Multiple Tasks: Manipulations of Task Difficulty and Training," ***International Journal of Psychophysiology***, Vol. 31, No. 2, pp. 129–145.

Galy, E., M. Cariou, and C. Mélan (2012), "What Is the Relationship between Mental Workload Factors and Cognitive Load Types?" ***International Journal of Psychophysiology***, Vol. 83, No. 3, pp. 269–275.

Greenwald, A. G. (1992), "New Look 3: Unconscious Cognition Reclaimed," ***American Psychologist***, Vol. 47, No. 6, p. 766.

Hart, S. G. (2006), "NASA-Task Load index (NASA-TLX); 20 Years Later," ***Proceedings of the Human Factors and Ergonomics Society Annual Meeting***, Vol. 50, No. 9, pp. 904–908.

Hart, S. G., and L. E. Staveland (1988), "Development of NASA-TLX (Task Load index): Results of Empirical and Theoretical Research," ***Advances in Psychology***, Vol. 52, pp. 139–183.

Hertzum, M., and K. D. Holmegaard (2013), "Perceived Time as a Measure of Mental Workload: Effects of Time Constraints and Task Success," ***International Journal of Human-Computer Interaction***, Vol. 29, No. 1, pp. 26–39.

Jaber, M. Y. (2011), **Learning Curves: Theory, Models, And Applications**, CRC Press, Boca Raton, FL.

Jaber, M., Z. Givi, and W. Neumann (2013), "Incorporating Human Fatigue and Recovery into the Learning – Forgetting Process," ***Applied Mathematical Modelling***, Vol. 37, No. 12, pp. 7287–7299.

Jaber, M. Y., and M. Bonney (2011). "The Lot Sizing Problem and the Learning Curve," in M. Y. Jaber, editor, **Learning Curves: Theory, Models, and Applications**, CRC Press, pp. 265–291.

Jung, H. S., and H. Jung (2001), "Establishment of Overall Workload Assessment Technique for Various Tasks and Workplaces," ***International Journal of Industrial Ergonomics***, Vol. 28, No. 6, pp. 341–353.

Kihlstrom, J. F. (1987), "The Cognitive Unconscious," ***Science***, Vol. 237, No. 4821, pp. 1445–1452.

Lindblom, J., and P. Thorvald (2014), "Towards a Framework for Reducing Cognitive Load in Manufacturing Personnel," ***Advances in Cognitive Engineering and Neuroergonomics***, Vol. 11, pp. 233–244.

Liu, Y., and C. Wickens (1988), "Patterns of Task Interference When Human Functions as a Controller or a Monitor," ***Proceedings of the 1988 IEEE International Conference on*** **Systems, Man, and Cybernetics, 1988**, Vol. 2, pp. 864–867.

Loft, S., P. Sanderson, A. Neal, and M. Mooij (2007), "Modeling and Predicting Mental Workload in En Route Air Traffic Control: Critical Review and Broader Implications," ***Human Factors***, Vol. 49, No. 3, pp. 376–399.

Marshall, S. P. (2002), "The Index of Cognitive Activity: Measuring Cognitive Workload," ***Proceedings of the 2002 IEEE 7th Conference on*** **Human Factors and Power Plants, 2002**, Scottsdale, AZ, USA, pp. 7-5–7-9.

May, J. G., R. S. Kennedy, M. C. Williams, W. P. Dunlap, and J. R. Brannan (1990), "Eye Movement Indices of Mental Workload," ***Acta Psychologica***, Vol. 75, No. 1, pp. 75–89.

Miller, S. (2001), "Workload Measures," *National Advanced Driving Simulator Guide,.Iowa City, Iowa, USA*

Mitchell, D. K. (2003), *Advanced Improved Performance Research Integration Tool (IMPRINT) Vetronics Technology Test Bed Model Development*, Report, Army Research Center, Adelphi, Maryland, USA

Murata, A. (2000), "Ergonomics and Cognitive Engineering for Robot-Human Cooperation," **Proceedings. 9th IEEE International Workshop on Robot and Human Interactive Communication, 2000.**, Rome, Italy, pp. 206–211.

Pesante, J. A., R. C. Williges, and J. C. Woldstad (2001), "The Effects of Multitasking on Quality Inspection in Advanced Manufacturing Systems," ***Human Factors and Ergonomics in Manufacturing & Service Industries***, Vol. 11, No. 4, pp. 287–298.

Reid, G. B., and T. E. Nygren (1988), "The Subjective Workload Assessment Technique: A Scaling Procedure for Measuring Mental Workload," ***Advances in Psychology***, Vol. 52, pp. 185–218.

Reinhard, M., R. Greifeneder, and M. Scharmach (2013), "Unconscious Processes Improve Lie Detection," ***Journal of Personality and Social Psychology***, Vol. 105, No. 5, p. 721.

Resnick, M. (2014), "Human Factors," in A. B. Badiru, editor, **Handbook of Industrial and Systems Engineering**, 2nd Edition, CRC Press, pp. 431–454.

Roscoe, A. H., and G. A. Ellis (1990). *A Subjective Rating Scale for Assessing Pilot Workload in Flight: A Decade of Practical use*, Technical Report TR 90019, Royal Aerospace Establishment, Farnborough, Hampshire, UK.

Rubio, S., E. Díaz, J. Martín, and J. M. Puente (2004), "Evaluation of Subjective Mental Workload: A Comparison of SWAT, NASA-TLX, and Workload Profile Methods," ***Applied Psychology***, Vol. 53, No. 1, pp. 61–86.

Stanton, N. A., A. Hedge, K. Brookhuis, E. Salas, and H. W. Hendrick, editors (2004), **Handbook of Human Factors and Ergonomics Methods**, CRC Press, Boca Raton, FL

Stanton, N. A., P. M. Salmon, G. H. Walker, C. Baber, and D. P. Jenkins (2010), **Human Factors Methods: A Practical Guide for Engineering and Design**, 1st Edition, Ashgate Publishing, Burlington, VT.

Tsang, P. S., and M. A. Vidulich (2006), "Mental Workload and Situation Awareness," in G. Salvendy (Ed.), Handbook of human factors and ergonomics (3rd ed., pp. 243–268).

Tsang, P. S., and V. L. Velazquez (1996), "Diagnosticity and Multidimensional Subjective Workload Ratings," ***Ergonomics***, Vol. 39, No. 3, pp. 358–381.

Vidulich, M. A., and P. S. Tsang (2012), "Mental Workload and Situation Awareness," in Gavriel Salvendy, editor, **Handbook of Human Factors and Ergonomics**, John Wiley, New York, NY, pp. 243–273.

Vidullch, M. A., G. F. Ward, and J. Schueren (1991), "Using the Subjective Workload Dominance (SWORD) Technique for Projective Workload Assessment," ***Human Factors: The Journal of the Human Factors and Ergonomics Society***, Vol. 33, No. 6, pp. 677–691.

Wickens, D. D. (1987), "The dual meanings of context: Implications for research, theory, and applications," in D. S. Gorfein & R. R. Hoffman, editors, *Memory and learning: The Ebbinghaus Centennial Conference* (pp. 135–152). Lawrence Erlbaum Associates, Inc.

Wickens, C., and C. Carswell (2006), in Gavriel Salvendy, editor, Handbook of Human Factors and Ergonomics, John Wiley, New York, NY

Wickens, C., and Holland (2000), Engineering Psychology and Human Performance.

Wickens, C. D. (2002), "Multiple Resources and Performance Prediction," ***Theoretical Issues in Ergonomics Science***, Vol. 3, No. 2, pp. 159–177.

Wickens, C. D. (2008), "Multiple Resources and Mental Workload," ***Human Factors***, Vol. 50, No. 3, pp. 449–455.

Wickens, C. D., J. G. Hollands, S. Banbury, and R. Parasuraman (2015), **Engineering Psychology & Human Performance.** Psychology Press.

Xie, B., and G. Salvendy (2000), "Prediction of Mental Workload in Single and Multiple Tasks Environments," ***International Journal of Cognitive Ergonomics***, Vol. 4, No. 3, pp. 213–242.

6 Workforce Time Management as a Factor of Productivity

Time is a critical factor in the adoption and utilization of technology. If we worship time, we will respect the timely impact of technology. Faluyi (2006) presents time as a factor of production. In a similar vein, Badiru (2006a) presents a poem on the passage of time. Benjamin Franklin, so long ago, presents a parody on time as the stuff of life.

> Dost, thou love life? Then do not squander time, for that is the stuff life is made of.
>
> **Benjamin Franklin**

POEM OF TIME

"On The Flight of Time.
What is the speed and direction of Time?
Time flies; but it has no wings.
Time goes fast; but it has no speed.
Where has time gone? But it has no destination.
Time goes here and there; but it has no direction.
Time has no embodiment. It neither flies, walks, nor goes anywhere.
Yet, the passage of time is constant."

Adedeji Badiru, 2006b

Faluyi (2006) uses an intriguing Nigerian storytelling style to convey the sense of the importance of time in production activities. He recounts: The young Nigerian had just arrived in Cork, Southern Ireland (Eire) from across the channel. After the usual pleasantries with the gentleman who came to meet him, the visitor asked how soon the day would start on the following day. With a glint in his eye, the Irish man replied, "There is an old Irish saying that when God made time, he made plenty of it. So, relax my friend, you are on this side of the channel now." The point being made was not lost on the visitor. Why, only some time earlier, the Duke of Edinburgh was visiting a city in the British Midlands. The City Elders were at the Railway station to welcome him. As the Duke alighted from the train, he went right up to the elders to apologize for coming late. Yes. His train had arrived TWO MINUTES late! That was the time when British Railways was under fire for not keeping to its schedule.

Imagine it, the Royal Train operated by British Railways arrived TWO MINUTES late. Such was the British attitude to time keeping and tardiness.

DOI: 10.1201/9781003537755-6

The Irish are also right. No one can contest that when God made time, he made plenty of it. As the God of Eternity, he is not subject to time constraints. However, the authentic account of creation shows that when He created time, He introduced seasons into it. In fact, the account of creation states, "And God went on to say let luminaries come to be in the expanse of the heavens to make division between the day and the night and they must serve as signs and for seasons and for days and years." [Gen. 2:14] So, from the beginning, the Creator intended for time to be measured, it was not fluid and unquantifiable. There were to be seasons, days, and years.

Rural communities all over the world appreciate that. There is a season for planting and a season for harvesting. Farmers know it and abide by it. No farmer plants and harvests with immediate effect. The farmer has a wholesome respect for time. For him, time is a factor of production, but the scale for measurement is much larger and has room for flexibility. A planting season or harvesting season does not have narrow, precise, and rigid timing limits for planting and harvesting. They have no need for hours and minutes.

ORIGIN OF HOURS AND MINUTES

In some of the earliest times, the Egyptians sought accuracy with respect to the use of time. First, the calendar year was established and then divided into 36 decans, each ten days long plus five extra days for a 365-day year. Each decan corresponded to a third of a zodiacal sign and was represented by a deanal constellation.

The Egyptians and later, the Babylonians (about 300–100 B.C.E), decided to divide the day from sunrise to sunset into twelve parts that are called hours. They also decided to divide the night, the time from sunset to sunrise, into twelve hours. But daylight and night are not the same length all year round. This system of measuring time was not very accurate because the actual length of the hour changed depending on the time of year. Someone finally figured out that by dividing the whole day into 24 hours (12 hours of the day plus 12 hours of the night), time could be measured much more accurately. This system is now universally used. This point is lost on Nigerians. We still consciously think of time in two divisions of twelve hours each and not a day of twenty-four hours. Thus, we speak of six hours in one division as 6:00 a.m. and six hours in the second division as 6:00 p.m. Some other communities speak of six hours and eighteen hours. Certainly, that conveys more accurately how much time of the day has been spent.

Despite this very early consciousness of time, time was not seriously considered as a factor of production. In calculating costs of production, time was not taken into consideration because no monetary value was ascribed to time.

In rural communities, north and south of the equator, time is seen largely in terms of seasons, namely spring, summer, autumn, and winter. It was long enough time to permit flexibility. The sub-division into day, hours, minutes, and seconds had no place. It was within that environment that the industrial revolution started in Europe. Economists and managers saw production only in terms of the material contributions to production, namely men, machines, money, and later, management. Time was no factor of production then.

It was not until the life of Fredrick Taylor, early in the 20th century, that time took its rightful place among the factors of production. Taylor, who was reputed as one of the fathers of Industrial Engineering, set out to study work as a subject. He sought to break every activity into its component parts, and it is the aggregate of those components that becomes the activity. That was the beginning of work study, which consists of time and motion study, a key aspect of any industrial engineering curriculum. Every motion took a precise amount of time, which can be measured and given a precise value. So, time now has a value, which can be managed. Today, man has moved on, not only to study and research materials management, but also time management. Time has come to be recognized as a factor of production, which must be properly managed. Amen!

Labor has an hourly rate of earnings and payments made in accordance with an agreed hourly rate of earning both for manual skills and professional skills. A consulting engineer would be paid in accordance with the hours he spends on the assignment, taking into consideration the level of proficiency required for the assignment. So, would an accountant as would a plumber, motor mechanic, or an electrician. Each would be paid in accordance with the hourly rate assigned to the complexity of his activity.

Interestingly, as the hourly rate for craftsmen rose steadily in Europe, the citizenry became restive and an increasing number of them sought to do the simpler craft tasks themselves. For example, a plumber was no longer called in to unblock kitchen sinks and later electricians were not invited to change fuses and light bulbs even on appliances. An age of Do It Yourself (DIY) was born. Today DIY has grown into a culture of maintenance that is sustained by specialized small tools and a regime of books and other materials that are specially provided for the untrained DO IT YOURSELF enthusiast. Nowadays, online YouTube guides are increasingly replacing printed DIY guidebooks. And that truly is what is required - enthusiasm. We talk glibly of a lack of maintenance culture in Nigeria. But truly, what has been done to promote such a culture? The environment that sustains such a culture is just not there. Who is there that wants to spare his time to have his equipment or machine opened up for diagnosis and repairs? Rather, the vast majority wants it replaced and we may as well start to accept that the replacement of machines or machine components is an acceptable mode of maintenance. There is a place for replacement maintenance within the context of maintenance management. And for us, literature and other materials, even spare parts are not available in Nigeria to facilitate repairs. So, we need to see a maintenance culture in another context, not necessarily a diagnosis of faults and repairs. All of these have direct and serious implications for workforce development and retention.

THE NIGERIAN TIME

In the context of workforce utilization, there is a pervasive Nigerian time. From the earliest of time, Nigerians have not seen time in terms of its smaller units of hours, minutes, and seconds. For the Nigerian workforce, the Irish saying is: "when God made time he made plenty of it." So, why have the constraint of hours and minutes? So, when a Nigerian says, "I'll see you tomorrow," he (or she) means, "I'll see you sometime during the daylight hours tomorrow." If he is pressed for a more definite time, then he says in the morning or in the afternoon. And that could mean any time before noon or the time the sun is at its highest point or any time after that before it goes completely dark. Time for the Nigerian workforce could be quite fluid.

CULTURAL DIVIDE ABOUT TIME

There is an anecdote that clearly illustrates this point. In the mid-1930s, the Resident of Oyo (an official of the British Government) administering the then Oyo province in today's southwestern Nigeria, had an appointment to see the Alafin of Oyo in the Resident's office at 10:00 a.m. Indeed, a few minutes before 10:00 a.m., the Resident was at the door, waiting to receive a highly respected and very senior ruler in Yoruba land. The Resident waited and waited. The Alafin arrived with his retinue of drummers, praise singers, and court jesters at about noon – two hours behind schedule. The Resident was furious, and he did little to hide his displeasure at what he considered uncivil and indeed bad manners. Imagine that - to keep the King of England's representative waiting for two hours! The Alafin was told that the Resident didn't take kindly to what he did; the appointment was for 10:00 a.m. and seeing what time the Alafin was coming in with such pomp and pageantry. The Alafin calmly told the interpreter to tell the Resident that it is when the Alafin arrives that it is 10:00 a.m. and not before. Yes, man is not bound by hours and minutes. It is what man calls it that it is. We are back at it again; when God made time, he made plenty of it.

Expatriates, who failed to understand this mental attitude of Nigerians in apportioning time, came to describe this approach as "African time." No, Africans don't have a time of their own. It is just that Africans, nay the Nigerians, do not operate within the precise context of hours, minutes, seconds. Even at this time of Nigerian history, how many meetings and events start at the scheduled time? Some people have sought to call attention to this failing. When they send out an invitation to a meeting or an event, they add the word prompt at the end of the time, such as 10:00 a.m. prompt. We have not come to understand what that word "prompt" means as the meeting or the event will start when it starts without any reference to the scheduled time. The military added their own dimension to the concept of time in Nigeria. It was the military that introduced "with immediate effect" or "effective immediately." For them, time has no impact on tasks. How, indeed, can anything be done with immediate effect? An official who is retired with immediate effect is required to write handing over notes! This requires some time to accomplish. A bridge (or was it a girder?) was to be completed "with immediate effect." Certainly, some actions are involved, and these actions involve some motions, which would take time. So, what then is the meaning of "with immediate effect?" That's an errant disrespect for time.

THE FALLACY OF PROJECT COMPLETION TIME

An example of this disrespect for time, with an unpleasant result, is shown in the construction of culverts and slabs. It is the norm to construct these culverts and slabs with concrete, sometimes reinforced concrete. Ordinarily, concrete takes twenty-eight days to cure. In other words, it will take about twenty-eight days for the mixture of stones, cement, sand, and water to acquire the strength and characteristics of concrete. But not so in Nigeria, the country cannot wait. We have no time. So, after a few days the culvert is put into use, or the slab is built upon. We cannot wait. Time has no value. The culvert fails or the slab collapses. We need to accept that time is a factor of production. Some time is required to carry out some tasks effectively, efficiently, and safely.

There is yet another angle to this indifferent attitude towards time. It is cost. A plumber or an electrician doing the same job in Ikoyi or Victoria Island would charge differently from what he would charge in Mushin or Ajegunle. Why? The craftsman does not have an hourly rate value. His charge is based on the ability of his client to pay. And that's another twist. Sometimes the craftsman may have to outsource materials and spare parts. He may have to go and find them and then come to collect money to buy them. Who pays for the time spent on that effort? The client only takes into consideration just the time spent on the assignment. So, how accurate is the contribution of this informal sector to the nation's Gross National Product?

BUDGET CYCLE CONUNDRUM

Let us look at it at another level. The national Budget and the State Government budgets are usually prepared for one year, January to December. In other words, the Government has determined what work it wants to do within the next twelve months and puts monetary value to it. However, it is sad that what is often of concern to most commentators, including Government officials involved in the preparation and approval of the budget, is the amount of money allocated and spent, not the actual work done. Of course, no one cares about the time required to finish the work assigned. Time is not a relevant factor. Otherwise, how can one explain the passing of the appropriation bill and signing it into law in March for a budget that would expire in December? So, the budget approval is to execute works scheduled for twelve months in nine months, at best. So, how much does the Government appreciate time as a factor of production?

No, Nigerians' appreciation of time must change. And the change is basic and must be at all levels. It seems good that current reform programs should be underpinned by a major shift in the attitude towards time.

NATIONAL TIME STANDARDS

A symbolic gesture may well be used to kick off the campaign to promote a change of attitude towards time. There are twenty-four hours in a day, but we treat it as twelve hours in that the measurement of time is within a twelve-hour framework. The counting starts at midnight, and when it gets to noon, we start counting from zero again. The Federal Government, by fiat or by an Act of the National Assembly, if that is the appropriate thing to do, should make our counting of time to be continuous from midnight to midnight. So, 1 p.m. that is one hour after noon becomes 13:00 hours and 5 p.m. becomes 17:00 hours and so forth. That is, indeed, the military time system. It will impress on all and sundry that we have spent seventeen hours out of the twenty-four hours we have in a day – not an illusory five hours. It is important that everyone should be conscious of the fact that time is a consumable product. Only twenty-four hours are available in a day. It cannot be extended, no matter how much one wishes or tries. So, a continuous reckoning of time may drive home that point, sub-consciously, but effectively. Comically, Badiru (2013, 2020, 2022) presents the secrets of the twenty-fifth hour in the day. This is not from a time creation fallacy, but from a time-saving and conservation approach.

GETTING A SHIFT ON TIMELINE

Government, by its pronouncements and actions, should also acknowledge that time is finite and show proper respect for it. So, a twelve-month budget that is for January to December should not be signed into law in March or whenever. The annual budget should be debated, approved, and signed into law no later than the last day of December of the preceding year. Thus, there will truly be an implementation period of twelve months for the approved budget. In that circumstance budgeting will make more sense. Those planning the budget will include only what they have the capacity to do within twelve months. Performance appraisal will be more realistic and meaningful. And how nice it would be when performance is appraised not in terms of money spent, but in terms of actual physical achievement.

Also, in giving assignments, realism should be considered in allocating time. If you ask for a thorough investigation of a major accident or a devastating disturbance and you ask the panel to submit its report within two weeks, then you are saying you do not want a thorough investigation or a good report. These things, if they are going to be acceptable, take time, and the Government must recognize that. Ridiculous timelines give the impression that the authority concerned is not aware of the volume of work involved or is not really concerned about doing a good job.

A consulting consortium was, once, requested to plan and design a Medical Diagnostic Center within three months. When the professionals protested that the time allowed was too short, they were asked if they wanted the job. "I suppose you are very busy now and you can't take the job," was suggested to them. The planning and design were completed in about four months, but the Diagnostic Center was never built and that was over twenty-five years ago. Some readers here, probably, have similar experiences. The government appears to think that money and manpower can shorten time. Not so! A popular parody in project management is to put more men on a job to get the job done.

This brings up a popular anecdote, in this regard, from that period when money seemed to be all that mattered. A renowned architect used to visit his construction sites on Saturdays. On many of those occasions, he took his son, John, along. Each week John listened to his father admonish the contractor that the job was slow and dragging. "Put more men on the job," Daddy would counsel. About this time, John's mother was pregnant. John saw mummy's tummy shooting out and he was told that mummy was making a baby inside her tummy. John waited for months to see mummy produce the baby she was making and nothing was forthcoming. So, John approached his father and said, "Daddy, why don't you admonish mummy to put more men on the job and accomplish this work of making a baby in time?" Can putting more men on the job speed up the making of a baby? Time is a factor of production, and it will be good for public officials to remember that when they are awarding construction contracts and, indeed, any contract, a wholesome respect for time should be considered as essential for the productive activity. "With immediate effect" is not a viable option for getting good results. As Kenneth Patton puts it, "By labour, we can find food and water, but all of our labour will not find for us another hour."

TIME EFFECT ON PRODUCTIVITY

Nigerians appear to be more concerned about the appearance of being busy rather than the substance of what is being achieved. Consider this example: A project has been identified. Promptly, a committee is set up and, of course, meetings are arranged. The public service is well known for this. Senior officers go from one meeting to another all day. They are very busy, attending meetings. When is the work done? Time is not often allowed for the implementation of the decisions made at such meetings. Time is not set aside for reviews of what is being done or for reflections on the decisions required and made. Consequently, time is not spent efficiently or effectively. Productivity is at its low ebb in such cases. This brings up Badiru's treatise of management meetings (Badiru, 1996), which is echoed below.

PERSPECTIVES ON WORKFORCE MEETINGS

Parkinson's Law says, "Work expands to fill the available time."

Deji Badiru's mutation says, "Talk expands to fill the available time."

Meetings represent one avenue for information flow and knowledge sharing for group decision-making. Effective management of meetings is an important skill for any administrator or manager. Unfortunately, meetings often degenerate to time-wasters; consequently, obstructing productivity and detracting from other essential functions. This is because most meetings are poorly organized, improperly managed, called at the wrong time, or even unnecessary. In some organizations, meetings are conducted as a matter of routine requirement rather than necessity. While meetings are essential for communication and decision-making, they accomplish nothing if not managed properly. A meeting of thirty people wasting only thirty minutes in effect wastes fifteen full hours of employee time. That much time, in any organization, amounts to thousands of dollars in lost time. It does not make sense to use a one-hour meeting to discuss a task that will take only five minutes to perform. This is analogous to hiring someone at $100,000 annual salary to manage an annual budget of $5,000. In 1993, a newspaper columnist/humorist, Dan L. Stewart, wrote about management meetings:

> "Management meetings are rapidly becoming this country's biggest growth industry. As nearly as I can determine, the working day of a typical middle manager consists of seven hours of meetings, plus lunch. Half a dozen years ago at my newspaper, we hired a new middle management editor with an impressive reputation. Unfortunately, I haven't met her yet. On her first day at work, she went into a meeting and has never come out."
>
> Stewart concludes his satire with "I'm expected to attend the next meeting. I'm not sure when it's scheduled exactly. I think they're having a meeting this afternoon about that."

In the past, when an employee had a request, he went to his boss, who would say "yes" or "no" right away. The whole process might take less than one minute out of the employee's day. Nowadays several hierarchies of meetings may be required to review the request. Thus, we may have a departmental meeting, a middle management staff meeting, upper management meeting, executive meeting, steering committee meeting, ad hoc committee meeting, an off-site retreat, and a meeting with outside consultants, all for the purpose of reviewing the simple request. Deji's perspectives about meetings are summarized below:

1. Don't call for a meeting, unless it is absolutely necessary.
2. Most of the information passed out at meetings can be more effectively disseminated through simple memos. Emails should be fully exploited to replace most meetings.
3. The point of diminishing returns for any meeting is determined by the number of people that are needed for the meeting. The larger the number of people at a meeting, the lower the meeting's productivity. The extra attendees only serve to generate unconstructive and conflicting ideas that only impede progress.
4. Not being invited to a meeting should be viewed as an indication of the high value placed on an individual's time within the organization.
5. Regularly scheduled meetings with specific time slots often become a forum for social assemblies, if not managed properly.
6. The satirical adjourn time of a meeting is equal to the scheduled start time plus five times the number of agenda items minus the actual start time. Mathematically, this is expressed as:

$$L = (T + 5N) - S$$

Where:

L = optimal length.
T = scheduled time.
N = number of agenda items.
S = meeting start-up time (i.e., actual call-to-order time).
$S \leq 5N$.

The subtraction of S in the expression is to encourage calling the meeting to order promptly. The larger the value of S, the less time is available for actual business. If $S = 0$, then total meeting time is allocated for conducting business. If $S = 5N$, then there will be no time left to hold the meeting. Since it is difficult to do away with all meetings, we must maximize the output of those that remain by using these guidelines:

1. Do pre-meeting homework.
   a. List topics to be discussed (agenda).
   b. Establish the desired outcome for each topic.
   c. Determine how the outcome will be verified.
   d. Determine who really needs to attend the meeting.
   e. Evaluate the suitability of meeting time and venue.
   f. Categorize meeting topics (e.g., announcements, important, urgent).
   g. Assign time duration to each topic.
   h. Verify that the meeting is really needed.
   i. Consider alternatives to the meeting (e.g., memo, telephone, electronic mail).

2. Circulate written agenda prior to the meeting.
3. Start meeting on time.
4. Review agenda at the beginning.
5. Get everyone involved; if necessary, employ direct questions and eye contact.
6. Keep to the agenda; do not add new items unless absolutely essential.
7. Be a facilitator for meeting discussions.
8. Quickly terminate conflicts that develop from routine discussions.
9. Re-direct irrelevant discussions back to the topic of the meeting.
10. Retain leadership and control of the meeting.
11. Recap the accomplishments of each topic before going to the next. Let those who have made commitments (e.g., promise to investigate certain issues) know what is expected of them.
12. Have empathy for those invited to meetings --- end meeting on time.

Regardless of the structure and expediency exercised in conducting meetings, the real work is done behind the scenes. Note that, as social animals, we sometimes need separate "social meetings" where we can have all the chips and drinks we want, and take all the time that we want.

One definition of a meeting is "the confusion of one man, multiplied by the number of people present at the meeting," in accordance with the formula promulgated above.

In many countries, the concept of productivity is at the heart of management resources. Not so in Nigeria. Here, it is enough that a task has been done; no one cares for how long it takes or how well it is done. Perhaps an hourly rate of payment will call attention to this. If a member of the industrial workforce were to know that his (or her) pay depends on how much he can achieve in the time allowed for him, perhaps, he will be more particular as to how he does his (or her) job. Wage negotiations would be more meaningful. Productivity can be measured and appropriately rewarded.

RIGHTSIZING FOR TIME FACTOR

Currently the talk is about right sizing of many public organizations. While some of these organizations maintain idle staff members, some staff members, in the same organization, are overworked and accumulate a lot of genuine overtime work. Overtime should not be viewed as a part of a normal day's work. Dr. Donald Wetmore observes that life consists of seven vital areas: health, family, financial, intellectual, social, professional, and spiritual. While it is not possible to spend time every day in each area, it is, however, important that, in the long run, spending a sufficient quantity and quality of time in each area would produce a balanced life. For example, a failure to take time now for preventive healthcare will, most likely, force one to take time later on for illness management. Similarly, ignoring the family now could cost a lot of time to re-establish family relationships later on. Overtime or overwork arrangements disrupt the balance in a person's life and it is not productive. It is a sure source of stress.

CONSTRUCTIVE TIME MANAGEMENT

The life balancing mentioned in the preceding section brings up another work balancing template offered by Badiru (2013), regarding dividing each 24-hour day into three segments of Regular Work, Home Life, and Sleep. Efficient time management is important for both corporate and personal success. One simple essence of time management at the corporate level is starting meetings at the scheduled time. Rarely do meetings start on time within the Nigerian context, but to be fair, this is also true in many other nations, even in some advanced economies. Everyone has come to accept that the meeting would not start on time. So, no one plans to be at the venue on time. They set out for the venue late and then blame the traffic for coming late. If all were time-conscious, less time would be lost at the commencement or start of an event. If twenty people were at a meeting or an event that started fifteen minutes late, a total of five productive hours would be lost. If it started thirty minutes late, a total of ten productive hours have been lost. If, indeed, we have an average hourly rate for those present, the value of the time lost, just for waiting to start, will be seen as what it truly is, a colossal waste of human resources.

If more people were to use the hourly rate for calculating their income, how much money is lost to traffic jams will have a greater impact. Often, these traffic jams are the result of indiscipline. A driver or two cannot wait, time must be abridged, or even eliminated, because each driver has something that is considered very important to do. So, drivers jump queues or seek to join the traffic at odd and dangerous points, thus making a complete mess of the traffic, which could take hours to clear up. A healthy respect for time and its value would prevent traffic mishaps, as all will keep in mind the cost of the result of a traffic jam.

PERSONAL TIME VERSUS CORPORATE TIME

At the personal level, time management is more challenging. It is a type of self-management. Different people may judge identical lengths of time quite differently. Time can "fly," that is, a long period of time can seem to go by very quickly or can seem to drag on, as one performs a boring task. In explaining his theory of relativity, Albert Einstein has been quoted as saying that although sitting next to a pretty girl for an hour feels like a minute, placing one's hand on a hot stove for a minute feels like an hour. Consequently, at the personal level time management requires an important shift in focus. Many people spend their days in a frenzy of activity but achieve very little because they are not concentrating on the right things.

There is a need to concentrate on results and not on being busy. Goals must be precisely set, priorities established, and planning of time written out. Such careful self-management would produce an efficient and effective use of time.

Yes, time is an important factor in production. Several books have been written on it, and courses and seminars have been delivered on time management. Computer software tools are available to satisfy the need to control employees' time and make it easier to coordinate work and increase accountability of the workforce.

With positive changes in the management of time, the workforce will come to appreciate time as a finite factor of production.

REFERENCES

Badiru, A. B. (1996), "***Perspectives on Meetings***," Essay on Management Practice, School of Industrial Engineering, University of Oklahoma, Norman, OK.

Badiru, A. B. (2006a), "**Poem on the Flight of Time**," Working Paper of Project Management, Department of Industrial Engineering, University of Tennessee, Knoxville, TN.

Badiru, A. B. (2006b), "**The Flight of Time – What Is the Speed and Direction of Time?**" Personal Collection of Life's Poems, Knoxville, TN

Badiru, D. (2013), **8-by-3 Paradigm for Time Management: Balancing Work, Home, and Leisure**, iUniverse, Bloomington, IN.

Badiru, D. (2020), **The Twenty-Fifth Hour: Secrets to Getting More Done Every Day**, iUniverse, Bloomington, IN.

Badiru, D. (2022), **Rapidity: Time Management on the Dot**, iUniverse, Bloomington, IN.

Faluyi, A. O. (March 16, 2006), **Time: A Factor of Production**, Academy Lecture, The Nigerian Academy of Engineering, University of Lagos, Lagos, Nigeria.

7 Making Local and Buying Local
Case Example of Domestic Product Patronage

INTRODUCTION

Education is learning what you didn't even know you didn't know.

Daniel J. Boorstin

Making local products and buying locally made products have direct implication for local workforce development, retention, and advancement. High-quality products are the basis for accomplishing this ideal. Quality operations work best when approached from a systems perspective to improve quality of outputs. This is particularly essential when an optimization framework is desired across national and geographical boundaries. A system is a collection of interrelated elements whose collective output, together, is higher than the mere sum of the individual outputs. For a systems approach to be practical, profitable, realizable, and optimizable, partnerships and collaborative frameworks must be instituted at the domestic level. Buying locally made products is the most dependable approach to improving product quality. This paper presents a systems framework to justify buying locally made products for the strategic goal of quality improvement. The DEJI Systems Model leads to a structured requirement for product design, evaluation, justification, and integration. In the premise of this paper, integration is regarding the domestic platform of the manufacturer. The methodology of this paper is, particularly, essential for developing economies.

It is a systems world. Systems thinking makes good things happen. The proverbial analysis paralysis is a constant threat in operations management. It is the premise of this paper that a structured systems approach can help advance product quality at the domestic level. The methodology of this paper, based on the trademarked DEJI Systems Model® (Badiru, 2023), can enforce a structure that permits a streamlined quality that can be more on target for product quality, from a systems perspective. Operations work best when approached from a systems perspective. This is particularly essential when an optimization framework is desired across national industrial development plans. (Badiru, 1993) In this regard, a system is a collection of interrelated elements whose collective output, together and in unison, is higher than the mere sum of the individual outputs. For a systems approach to be practical, profitable, realizable, and optimizable, partnerships and collaborative frameworks must

DOI: 10.1201/9781003537755-7

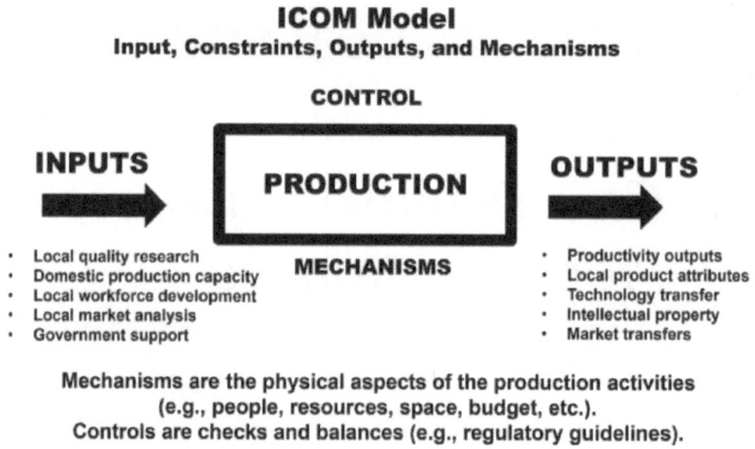

FIGURE 7.1 Quality systems inputs-process-outputs relationships.

be instituted, subject to real-world constraints involving the people aspects of the systems. **Systems make things possible, but people make things happen.** If we desire to improve product quality in the final analysis, we must integrate people's considerations into the manufacturing equation. This is precisely what the DEJI Systems Model advocates for in the implementation phase of any system.

The advocacy presented in this paper is to push efforts to buy locally made products, particularly in the developing countries, where the cultural and social norms may typically prefer buying imported products. The fact is that if locally made products are not embraced, the local manufacturers may not have the incentive to invest in quality-improvement processes. As the proverbial saying goes, "Rome was not built in a Day." So, high-quality products cannot be achieved overnight. The highest levels of quality observed in any consumer product come only after gradual improvement iterations, using the research-and-development framework of input-process-output. The ICON model presented in Figure 7.1 illustrates how what is put in determines what the manufacturers get as outputs. Market incentives help the manufacturer to put in more in terms of inputs, so that the production process can be enhanced, towards the eventual goal of better products. The ICOM Model facilitates the conversion or transformation of input into output, under a set of constraints (controls) and utilizing a set of mechanisms (people, tools, and process) to manifest the productive capabilities. Humans in the loop are expected to serve as system integrators.

Quality Basis for Industrial Development

Every manufacturer aspires to compete globally through product acceptance in the marketplace. Acceptance is predicated on acceptable product quality. This onus is heavier on manufacturers in developing countries. However, manufacturers in developed economies also desire to stay ahead of their competition through world-class quality accomplishments. Thus, there is persistent pressure for additional research and development in both new and established products. National industrial

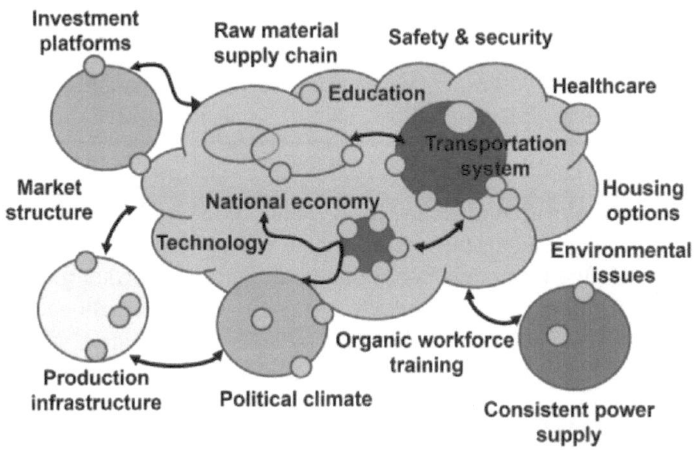

FIGURE 7.2 Interweaving domestic factors affecting product quality.

development plans around the world often contain statements about product quality. As long ago as the 18th century, Adam Smith (1776) opined that product quality "renders them fit to be the instruments of commerce and circulation." Poor quality will stymy product circulation in the long run. Figure 7.2 illustrates the interweaving systems elements that influence product quality and national industrial development. For example, in the developing country of Nigeria, several elements interface to cause disruption of the product quality infrastructure, despite the government's bold attempts to remedy industrial production. In 2000, the former President of Nigeria presented an economic direction for the country in a paper titled, "Making a Quantum Leap in Nigeria's Economic Development," ***NESG Economic Indicators***, Vol. 6, No. 3, September 2000, pp. 42–46. (NESG: Nigeria's Economic Summit Group, Lagos, Nigeria) Some of the key statements in the paper are echoed below, with this paper's assessment comments, in the context of national quality pursuit.

- Worldwide, economic success is often seen as one of the most important measures of the quality of governance. In this context, quality of national governance can directly affect the quality of products and services emanating from the nation.
- It is generally agreed that Nigeria's economic progress over the years has neither matched its enormous resource endowment nor the expectations of the citizens. Unfortunately, citizens have become disillusioned over the years.
- Nigeria's economic development must be anchored on government focus on creating the enabling environment for the private sector to thrive rather than struggling with the production of domestic goods and services. This nonlinear correlation is yet to unfold.
- The macroeconomic environment is more stable today than in the past. This is debatable in present-day Nigeria.
- Institutions for private sector development, including the creation of appropriate legal framework, must rank highly in the list of priorities of the

administration. Unfortunately, this stated aspiration has not panned out since that 2000 platform.

- Inflation, in the 2000 example, has stabilized and was judged to be the lowest in the years preceding 2000. This is a sharp contrast to the reality of today, in 2024.
- Implementation is the Achilles' heel of Nigeria's national effort, and until the country gets around to it, success cannot be assured. It seems this country is still in circuitous pursuit of the aspired implementation.
- Success on the economic front remains the most important guarantee for the sustainability of the democratic system. Yes, indeed, national economic success paves the way for democratic pursuits in a nation. A vibrant economy is the lynchpin for democracy and sustainable production of quality.

All the points stated in the 2000 national statement are well noted, but a stable realization is still far off. It is the argument of this paper that a structured systems-based approach is required for success to be realized. Rhetoric and aspirations alone cannot achieve quality goals in a competitive global market. It is on this basis that this paper presents the DEJI Systems Model for product quality design, evaluation, justification, and integration. **Design**, in the context of product quality, generically covers several objectives, including planning, conceptualization, organizing, brainstorming, negotiation, and so on. The systems approach of **Design** leads to a structured requirement to conduct a formal **Evaluation**, followed by a rigorous **Justification**, followed by, above all, **Integration**. Programs often fail due to a lack of sustainable integration of efforts. The **DEJI Systems Model** forces a structured and rewarding approach to what could, otherwise, become disjointed efforts. With the structured methodology of this paper, quality measuring and monitoring can be better undertaken. (Bendell et al, 1993)

Industry-Government Collaboration

Figure 7.3 presents the iconic representation of the DEJI Systems Model within a collaborative model that encompasses Industry, Government, and University. As a research platform, the university is charged with the creation of new knowledge and the advancement of existing knowledge, for the benefit of industry. The government's role is to facilitate the transmission of university research into practical tools for industry. As presented in the figure, coordination and integration as well as communication and collaboration serve as anchors for embracing locally made products.

In the integration context and the theme of product quality, a system is a group of manufacturers working together towards some common objective or purpose, which fits the goal of product advancement. Any system of production is composed of components, attributes, and relationships that constitute the core of how the system interacts with its environment.

Case Example from Nigeria

Quality research, modeling, and optimization can be more effective if the problem formulation encompasses the framework of factors, attributes, and relationships related to the domestic production capacity and local market conditions. Street-level

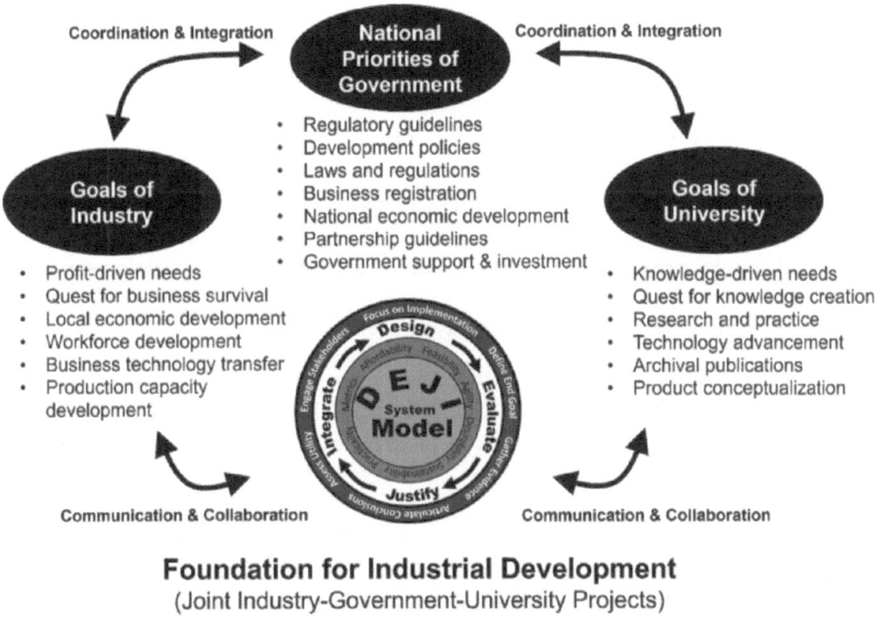

FIGURE 7.3 DEJI systems model for industry-government collaboration.

consumers (buyers) are a key part of the overall system of quality advancement. If ordinary consumers do not embrace and patronize locally made products, the path to national industrial development will be curtailed. Thus, grassroot efforts must be made to cultivate market acceptance of locally made products, even if it is on a platform of national patriotism and collective commitment to national advancement. A case in point, those seeking to purchase vehicles in Nigeria tend to opt for imported vehicles, particularly Hondas and Toyotas. These are claimed to be of a high quality and are made for the bumpy Nigerian roads. At the same time, an emerging local auto maker, Innoson Vehicle Manufacturing Co. Ltd, is available to satisfy market needs. The problem is that buyers need to be convinced to give the local manufacturer a chance. Proverbially, to convince the local buyer. In its own product advertisements, Innoson Vehicle Manufacturing (IVM) touts the following benefits:

- Locally enmeshed with local market needs.
- Over thirteen years of relentless pursuit of perfection (That is, better quality, in this case).
- Prioritization of customer protection and satisfaction.
- Legacy of engagement with cultural heritage. This is the type of *integration* that DEJI Systems Model advocates.

As an arm of domestic competition in Nigeria, the Lagos State Government partners with an automobile manufacturer in China. As conveyed by the basic definition of a system, elements that partner and work together can yield greater results and impacts than the sum of the elements. In this case study, the state-level government entity collaborates with GAC (Guangzhou Automobile Group Company: www.gacmotor.com.ng) in China to produce the GSx series of vehicles. When constructive competition exists, the drive for higher quality develops within each manufacturing enterprise. In this expanded scope of the Nigerian auto market, the traditional sole source of vehicle purchase (imported market) is challenged by two local producers: Innoson and GAC. With three competing purchase options, local consumers will gradually be introduced to the performance capabilities of locally made vehicles. Over time, the locally made vehicles will gain traction in the market, may eventually outstrip their imported counterparts. This is the foundation for sustainable industrial development.

Enhancing Domestic Product Quality

Modeling, in the context of quality management, is predicated on data availability, acquisition, storage, retrieval, and analytical usage. (Badiru, 2020) Sink et al (1995) highlight quality and productivity improvement pathways. They provide guidelines for developing the knowledge and skills to lead large-scale quality improvement efforts. Peacock and Badiru (2023) also present tools and techniques for industrial engineering in quality systems design. The guidelines provided a center for using data (through market feedback) to design new pathways for quality improvement. In the sense of what a quality manager might be interested in, a Dilbert comic says it best:

> You don't go to war with the data you need. You go to war with the data you have.
>
> **Dilbert Comic, May 7, 2020**

In a similar vein, we have the following famous quote:

> You go to war with the army you have, not the army you might want or wish to have at a later time.
>
> **Donald Rumsfeld**

Several mathematical models and variations of data analytics are used in practice for quality modeling and improvement. Below are some of the more common models:

1. Conceptual models.
2. Theoretical models.
3. Observational models.
4. Experimental models.
5. Empirical models.
6. Iconic models.
7. A-priori reasoning.

For the purpose of integrating production capability with the prevailing local market needs, presented below are pertinent questions in the pursuit of product integration:

- What are the unique characteristics of each component in the integrated response system?
- How do the characteristics complement one another?
- What physical interfaces exist among the resilience components?

- What data and/or information interfaces exist among the resilience components?
- What ideological differences exist among the humans involved in resilience efforts?
- What are the data flow requirements for the resilience components?
- What internal and external factors are expected to influence the integrated system?
- What are the relative priorities assigned to each component of the integrated system?
- What are the strengths and weaknesses of the integrated system?
- What resources are needed to keep the integrated resilience system operating satisfactorily?
- Which organizational unit has primary responsibility for the integrated resilience system?

In this regard, the benefits of a structured systems template, such as the DEJI Systems Model, are that simplicity, acceptability, relevance, and repeatability can be achieved in the quantitative quality modeling process. A scaled optimization model at one level can pave the way for a higher inclusive model at the next analytical level. Several optimization models published in literature provide a justification and a pathway for scalable optimization for industrial production. With a focus on energy optimization, Singh and Lou (2006) present a hierarchical Pareto optimization model for sustainable development of industrial ecosystems. Hsu et al (2012) presents heuristics, optimization, and equilibrium analysis for automated wargames in a way that mixes qualitative and quantitative parameters in the global model, considering political, military, economic, social, infrastructure, and information realities of each participating entity. Using a computer simulation of the model, optimal paths were identified in the wargames. In process optimization research, Al-Quradaghi et al (2020) present a generalized framework for the design of eco-industrial parks as a case study of end-of-life vehicles, in which the end of life is predicated on the initial product quality. In a single-loop deterministic method (SLDM) for reliability-based quality design optimization (RBDO) research, Wu et al (2012) introduce a quantitative problem formulation, which can be adapted for generalized quality-improvement modeling. Odior et al (2010) introduce a novel approach for determining feasible solutions of a multicriteria assignment problem, such as we may find in the deployment and assignment of quality attributes to a complex product configuration. The assignment of product attributes is akin to the allocation of limited resources (inputs) to the overall composition of the product, which can be modeled as a knapsack problem, with the generic pictorial representation shown in Figure 7.4. The expression at the top of the sack represents a quantitative summation of the attributes (weights) added to the sack, subject to the weight limitation or capacity (W) of the sack. The knapsack formulation is solvable mathematically by the general integer linear programming (ILP) algorithms. The formulation can be further complicated by including specific time windows of product operation. The increasing complexity of formulation makes it imperative that we consider a systems-based restructuring of quality problems to facilitate scalable optimization that can be implemented and integrated hierarchically and incrementally. This is where the partnership with a

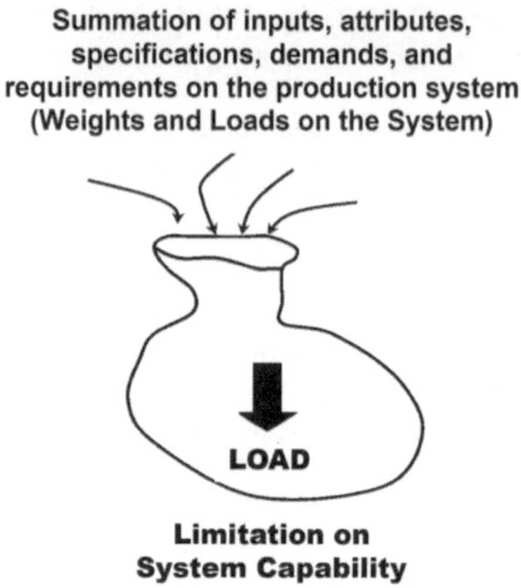

FIGURE 7.4 Representation of putting loads on capacity-limited production.

university research program is an essential component of a nation's industrial development pursuits. The more we demand from a product, the more inputs we must commit to the product. In some high-quality products, the products may be over-designed for higher customer satisfaction when a basic version of the product may suffice for normal usage. This is why some vehicles are produced with designations of SE, LE, XL, XLT, Denali, and so on to convey the level of options "loaded" onto the product. Table 7.1 provides a brief summary of the scope and structure of the DEJI Systems Model, applied to product attributes and characteristics.

The conventional Pareto analysis can play a role in certain aspects of actualizing the DEJI Systems Model. Some practical realities and expectations to guide a Pareto optimization include:

- 80 percent of the value generated comes from 20 percent of the assets.
- 80 percent of problems originate from 20 percent of the population.
- 20 percent of beneficiaries consume 80 percent of the available resources.
- 20 percent of participants contribute 80 percent of the required resources.

TABLE 7.1
Optimization Stages and Compartments of DEJI Systems Model

Design	Evaluate	Justify	Integrate
Specify requirements.	Identify attributes.	Make a case.	• Ensure implementation.
• Physical	• Cost	• Desirability	• Alignment
• Conceptual	• Schedule	• Acceptability	• Sustainability
• Expectations	• Quality	• Etc.	• Etc.
• Etc.	• Etc.		

Communicating Domestic Product Quality

The local manufacturers must "make noise" about their products in terms of aggressive marketing to communicate the quality of their products. For local manufacturers, word-of-mouth advertisement alone is not sufficient. It must be buttressed with aggressive marketing campaigns. Even in the best of systems, communication, cooperation, and coordination are essential for promoting, executing, and sustaining quality operations. Every initiative of a manufacturer is a "project" and should be managed as such. Each project is predicated on the interplay of people, technology, and process. As the qualitative part of operational research on project management, Badiru (2008) introduces the Triple C approach to project management to facilitate communication, cooperation, and coordination of project efforts. This is essential because the people aspects of implementing an optimized model require human-centric communication, cooperation, and coordination. This feeds into the justification and integration stages of the DEJI Systems Model as compartmentalized in Table 7.1. For example, the formulation of constraints faced in market integration and acceptance could involve the following:

- Cultural and social compatibility issues.
- Administrative and management compatibility.
- Physical and technical compatibility.
- Maintenance and adaptation interface.
- Modular product characteristics.
- Pilot or phased integration.

RECOMMENDATIONS

As Badiru and Tourangeau (2023) emphasize, leadership matters whether we are working with a team of quantitative analysts or a group of seat-of-the-pants decision makers. Leveraging a systems approach to analytical modeling for quality management is a viable path forward in our evolving highly contested global markets. The DEJI Systems Model and other similarly structured tools create a leadership avenue for new approaches to quality improvement initiatives. A systems approach makes things possible, but people, in terms of human leadership, make things happen. Any industrial development program requires leveraging innovation. However, assessing innovation is essential (Badiru & Tourangeau, 2024) to ensure that system metrics, rubrics, and standards are considered in developing competitive product quality at the local production level. Further applied research is recommended for exploring applications of not only the DEJI Systems Model but also other systems-based methodologies also.

REFERENCES

Al-Quradaghi, S., Q. P. Zheng, and A. Eklamel (2020), "Generalized Framework for the Design of Eco-Industrial Parks: Case Study of End-of-Life Vehicles," ***Sustainability 2020***, Vol. 12, pp. 6612. doi:10.3390/su12166612.

Badiru, A. B. (1993), **Managing Industrial Development Projects: A Project Management Approach**, Van Nostrand/John Wiley, New York, NY.

Badiru, A. B. (2008), **Triple C Model of Project Management: Communication, Cooperation, and Coordination**, Taylor & Francis CRC Press, Boca Raton, FL.

Badiru, A. B. (2020), **Data Analytics: Handbook of Formulas and Techniques**, CRC Press/Taylor and Francis, Boca Raton, FL.

Badiru, A. B. (2023) **Systems Engineering Using DEJI Systems Model: Design, Evaluation, Justification, and Integration With Case Studies and Applications**, Taylor & Francis CRC Press, Boca Raton, FL.

Badiru, A.B., and M. Tourangeau (2024), **Assessing Innovation: Metrics, Rubrics, and Standards**, CRC Press/Taylor and Francis, Boca Raton, FL.

Badiru, A.B., and M. L. Tourangeau (2023), **Leadership Matters: Industrial Engineering Framework for Developing and Sustaining Industry**, Taylor & Francis CRC Press, Boca Raton, FL.

Bendell, T., J. Kelly, T. Merry, and F. Sims (1993), **Quality: Measuring and Monitoring**, Century Business, London.

Hsu, W., E. A. Newell, J. Zhuang, and D. O. Ross (2012), "Heuristics, Optimization, and Equilibrium Analysis for Automated Wargames," in ***Proceedings of the 2012 Industrial and Systems Engineering Research Conference***, G. Lim and J. W. Herrmann, editors, Orlando, FL.

Odior, A. O., O. E. Charles-Owaba, and F. A. Oyawale (March 2010), "Determining Feasible Solutions of a Multicriteria Assignment Problem," ***Journal of Applied Science and Environmental Management***, Vol. 14, No. 1, pp. 35–38.

Peacock, B., and A. Badiru (2023), **Industrial Engineering in Systems Design: Guidelines, Practical Examples, Tools, and Techniques**, Taylor & Francis CRC Press, Boca Raton, FL.

Singh, A., and H. H. Lou (2006), "Hierarchical Pareto Optimization for the Sustainable Development of Industrial Ecosystems," ***Industrial and Engineering Chemistry Research***, Vol. 45, pp. 3265–3279.

Sink, D., W. T. Scott, C. S. Morris, and Johnston (1995), **By What Methods?** IIE Press, Norcross, GA.

Smith, A. (1776), **The Wealth of Nations**, Meta Libri, W. Strahan and T. Cadell Publishers, London.

Wu, T., M. Hu, A. B. Badiru, and Soni (2012), "A Single Loop Deterministic Method for Reliability Based Design Optimization," ***Engineering Optimization***, Vol. 45, No. 4, pp. 435–458.

8 Workforce and Technology Transfer Strategies

Technological progress is like an axe in the hands of a pathological criminal.

Albert Einstein

The opening quote reminds us that technology can easily be misused, if not properly controlled. Technology evolves for beneficial purposes, but its use can be misapplied, mis-transferred, or misunderstood. Using appropriate transfer strategies can ensure that manufacturing technology can avoid the pathological criminal analogy conveyed in Albert Einstein's quote.

Now that we have addressed the what, why, when, where, and when aspects of manufacturing technology in the preceding chapters, we now address the "how" of transferring manufacturing technology from one endpoint to another. Technology transfer is not just about the hardware components of technology. It can involve a combination of several components, including software (computer-based) and "skin ware" (people-based). Thus, this chapter addresses the transfer of knowledge as well as the transfer of skills.

Why reinvent the wheel when it can be transferred and adopted from existing wheeled applications? The concepts of project management can be very helpful in planning for the adoption and implementation of new industrial technology. Due to its many interfaces, the area of industrial technology adoption and implementation is a prime candidate for the application of project planning and control techniques. Technology managers, engineers, and analysts should try to take advantage of the effectiveness of project management tools. This applies the various project management techniques that have been discussed in the preceding chapters to the problem of industrial technology transfer. The project management approach is presented within the context of technology adoption and implementation for industrial development. Project management guidelines are presented for industrial technology management. The Triple C model of Communication, Cooperation, and Coordination is applied as an effective tool for ensuring the acceptance of new technology. The importance of new technologies in improving product quality and operational productivity is also discussed. This chapter also outlines the strategies for project planning and control in complex technology-based operations.

DOI: 10.1201/9781003537755-8

CHARACTERISTICS OF TECHNOLOGY TRANSFER

To transfer technology, we must know what constitutes technology. A working definition of technology will enable us to determine how best to transfer it. A basic question that should be asked is:

What is technology?

Technology can be defined as follows:

Technology is a combination of physical and nonphysical processes that make use of the latest available knowledge to achieve business, service, or production goals.

Technology is a specialized body of knowledge that can be applied to achieve a mission or purpose. The knowledge concerned could be in the form of methods, processes, techniques, tools, machines, materials, and procedures. Technology design, development, and effective use are driven by effective utilization of human resources and effective management systems. Technological progress is the result obtained when the provision of technology is used in an effective and efficient manner to improve productivity, reduce waste, improve human satisfaction, and raise the quality of life.

Technology all by itself is useless. However, when the right technology is put to the right use with an effective supporting management system, it can be very effective in achieving industrialization goals. Technology implementation starts with an idea and ends with a productive industrial process. Technological progress is said to have occurred when the outputs of technology in the form of information, instrument, or knowledge that is used productively and effectively in industrial operations leads to lowering costs of production, better product quality, higher levels of output (from the same amount of inputs), and higher market share. The information and knowledge involved in technological progress includes those which improve the performance of management, labor, and the total resources expended for a given activity.

Technological progress plays a vital role in improving overall national productivity. Experience in developed countries such as the United States shows that in the period 1870–1957, ninety percent of the rise in real output per man-hour can be attributed to technological progress. It is conceivable that a higher proportion of increases in per capita income is accounted for by technological change. Changes occur through improvements in efficiency in the use of existing technology. That is, through learning and the adaptation of other technologies, some of which may involve different collections of technological equipment. The challenge to developing countries is how to develop infrastructure that promotes, uses, adapts, and advances technological knowledge.

Most of the developing nations today face serious challenges, arising not only from the world-wide imbalance of dwindling revenue from industrial products and oil, but also from major changes in a world economy that is characterized by competition, imports, and exports of not only oil, but also of basic technology, weapon systems, and electronics. If technology utilization is not given the right attention in all sectors of the national economy, the much-desired industrial development cannot occur or cannot be sustained. The ability of a nation to compete in the world market will, consequently, be stymied.

TABLE 8.1
The "ilities" of Manufacturing Technology

Characteristics	Definitions, Questions, and Implications
Adaptability	Can the technology be adapted to fit the needs of the organization? Can the organization adapt to the requirements of the technology?
Affordability	Can the organization afford the technology in terms of first cost, installation cost, sustainment cost, and other incidentals?
Capability	What are the capabilities of the technology with respect to what the organization needs? Can the technology meet the current and emerging needs of the organization?
Compatibility	Is the technology compatible with existing software and hardware?
Configurability	Can the technology be configured for the existing physical infrastructure available within the organization?
Dependability	Is the technology dependable enough to produce the outputs expected?
Desirability	Is the particular technology desirable for the prevailing operating environment of the organization? Are there environmental issues and/or social concerns related to the technology?
Expandability	Can the technology be expanded to fit the changing needs of the organization?
Flexibility	Does the technology have flexible characteristics to accomplish alternate production requirements?
Interchangeability	Can the technology be interchanged with currently available tools and equipment in the organization? In case of operational problems, can the technology be interchanged with something else?
Maintainability	Does the organization have the wherewithal to maintain the technology?
Manageability	Does the organization have adequate management infrastructure to acquire and use the technology?
Re-configurability	When operating conditions change or organizational infrastructure change, can the technology be re-configured to meet new needs?
Reliability	Is the technology reliable in terms of technical, physical, and /or scientific characteristics?
Stability	Is the technology mature and stable enough to warrant an investment within the current operating scenario?
Sustainability	Is the organization committed enough to sustain the technology for the long haul? Is the design of the technology sound and proven to be sustainable?
Volatility	Is the technology devoid of volatile developments? Is the source of the technology devoid of political upheavals and/or social unrests?

The important characteristics or attributes of new technology may include productivity improvement, improved quality, cost savings, flexibility, reliability, and safety. An integrated evaluation must be performed to ensure that a proposed technology is justified both economically and technically. The scope and goals of the proposed technology must be established right from the beginning of the project. Table 8.1 summarizes some of the common characteristics of technology transfer for a well-rounded assessment.

An assessment of a technology transfer opportunity will entail a comparison of departmental objectives with overall organizational goals in the following areas:

1. **Industrial Marketing Strategy:** This should identify the customers of the proposed technology. It should also address items such as market cost of proposed product, assessment of competition, and market share. Import and export considerations should be a key component of the marketing strategy.
2. **Industry Growth and Long-range Expectations:** This should address short-range expectations, long-range expectations, future competitiveness, future capability, and prevailing size and strength of the industry that will use the proposed technology.
3. **National Benefit:** Any prospective technology must be evaluated in terms of direct and indirect benefits to be generated by the technology. These may include product price versus value, increase in international trade, improved standard of living, cleaner environment, safer workplace, and higher productivity.
4. **Economic Feasibility:** An analysis of how the technology will contribute to profitability should consider past performance of the technology, incremental benefits of new technology versus conventional technology, and value added by the new technology.
5. **Capital Investment:** Comprehensive economic analysis should play a significant role in the technology assessment process. This may cover an evaluation of fixed and sunk costs, cost of obsolescence, maintenance requirements, recurring costs, installation cost, space requirement cost, capital substitution options, return on investment, tax implications, cost of capital, and other concurrent projects.
6. **Resource Requirements:** The utilization of resources (human resources and equipment) in the pre-technology and post-technology phases of industrialization should be assessed. This may be based on material input-output flows, high value of equipment versus productivity improvement, required inputs for the technology, expected output of the technology, and utilization of technical and non-technical personnel.
7. **Technology Stability:** Uncertainty is a reality in technology adoption efforts. Uncertainty will need to be assessed for the initial investment, return on investment, payback period, public reactions, environmental impact, and volatility of the technology.
8. **National Productivity Improvement:** An analysis of how the technology may contribute to national productivity may be verified by studying industrial throughput, efficiency of production processes, utilization of raw materials, equipment maintenance, absenteeism, learning rate, and design-to-production cycle.

EMERGENCE OF NEW TECHNOLOGY

New industrial and service technologies have been gaining more attention in recent years. This is due to the high rate at which new productivity improvement technologies are being developed. The fast pace of new technologies has created difficult implementation and management problems for many organizations. New technology can be successfully implemented only if it is viewed as a system whose various components must be evaluated within an integrated managerial framework. Such a framework is provided by a project management approach. A multitude of new technologies have emerged in recent years. It is important to consider the peculiar characteristics of new technology before establishing adoption and implementation strategies. The justification for the adoption of new technology is usually a combination of several factors rather than a single characteristic of the technology. The potential of a specific technology to contribute to industrial development goals must be carefully assessed. The technology assessment process should explicitly address the following questions:

What is expected from the new technology?
Where and when will the new technology be used? How is the new technology similar to or different from existing technologies?
What is the availability of technical personnel to support the new technology?
What administrative support is needed for the new technology?
Who will use the new technology?
How will the new technology be used?
Why is the technology needed?

The development, transfer, adoption, utilization, and management of technology is a problem that is faced in one form or another by business, industry, and government establishments. Some of the specific problems in technology transfer and management include the following:

- Controlling technological change.
- Integrating technology objectives.
- Shortening the technology transfer time.
- Identifying a suitable target for technology transfer.
- Coordinating the research and implementation interface.
- Formal assessment of current and proposed technologies.
- Developing accurate performance measures for technology.
- Determining the scope or boundary of technology transfer.
- Managing the process of entering or exiting a technology.
- Understanding the specific capability of a chosen technology.
- Estimating the risk and capital requirements of a technology.

Integrated managerial efforts should be directed at the solution of the problems stated above. A managerial revolution is necessary to cope with the ongoing technological revolution. The revolution can be initiated by modernizing the long-standing and obsolete management culture relating to technology transfer. Some of the managerial functions that will need to be addressed when developing a technology transfer strategy include the following:

1. Development of a technology transfer plan.
2. Assessment of technological risk.

3. Assignment/reassignment of personnel to implement the technology transfer.
4. Establishment of a transfer manager and a technology transfer office. In many cases, transfer failures occur because no individual has been given the responsibility to ensure the success of technology transfer.
5. Identification and allocation of the resources required for technology transfer.
6. Setting of guidelines for technology transfer. For example,
   a. Specification of phases (Development, Testing, Transfer, etc.).
   b. Specification of requirements for inter-phase coordination.
   c. Identification of training requirements.
   d. Establishment and implementation of performance measurement.
7. Identify key factors (both qualitative and quantitative) associated with technology transfer and management.
8. Investigate how the factors interact and develop the hierarchy of importance for the factors.
9. Formulate a loop system model that considers the forward and backward chains of actions necessary to effectively transfer and manage a given technology.
10. Track the outcome of the technology transfer.

Technological developments in many industries appear in scattered, narrow, and isolated areas within a few selected fields. This makes technology efforts rarely coordinated, thereby hampering the benefits of technology. The optimization of technology utilization is, thus, very difficult. To overcome this problem and establish the basis for effective technology transfer and management, an integrated approach must be followed. An integrated approach will be applicable to technology transfer between any two organizations, whether public or private.

Some nations concentrate on the acquisition of bigger, better, and faster technology, but little attention is given to how to manage and coordinate the operations of the technology once it arrives. When technology fails, it is not necessarily because the technology is deficient. Rather, it is often the communication, cooperation, and coordination functions of technology management that are deficient. Technology encompasses factors and attributes beyond mere hardware, software, and "skin ware," which refers to human issues affecting the utilization of technology. This may involve social-economic and cultural issues of using certain technologies. Consequently, technology transfer involves more than the physical transfer of hardware and software. Several flaws exist in the common practices of technology transfer and management. These flaws include the following:

- **Poor fit:** This relates to an inadequate assessment of the needs of the organization receiving the technology. The target of the transfer may not have the capability to properly absorb the technology.
- **Premature transfer of technology:** This is particularly acute for emerging technologies that are prone to frequent developmental changes.
- **Lack of focus:** In the attempt to get a bigger share of the market or gain early lead in the technological race, organizations frequently force technology in many incompatible **directions.**

- **Intractable implementation problems:** Once a new technology is in place, it may be difficult to locate sources of problems that have their roots in the technology transfer phase itself.
- **Lack of transfer precedents:** Very few precedents are available on the management of brand-new technology. Managers are, thus, often unprepared for their new technology management responsibilities.
- **Stuck on Technology:** Unworkable technologies sometimes continue to be recycled needlessly in the attempt to find the "right" usage.
- **Lack of foresight:** Due to the nonexistence of a technology transfer model, managers may not have a basis against which they can evaluate future expectations.
- **Insensitivity to external events:** Some external events that may affect the success of technology transfer include trade barriers, taxes, and political changes.
- **Improper allocation of resources:** There aren't usually enough resources available to allocate to technology alternatives. Thus, a technology transfer priority must be developed.

The following steps provide a specific guideline for pursuing the implementation of manufacturing technology transfer:

1. Find a suitable application.
2. Commit to an appropriate technology.
3. Perform economic justification.
4. Secure management support for the chosen technology.
5. Design the technology implementation to be compatible with existing operations.
6. Formulate a project management approach to be used.
7. Prepare the receiving organization for the technology change.
8. Install the technology.
9. Maintain the technology.
10. Periodically review the performance of the technology based on prevailing goals.

TECHNOLOGY TRANSFER MODES

The transfer of technology can be achieved in various forms. Project management provides an effective means of ensuring the proper transfer of technology. Three technology transfer modes are presented here to illustrate basic strategies for getting one technological product from one point (technology source) to another point (technology sink). A conceptual integrated model of the interaction between the technology source and sink is presented in Figure 8.1.

The university-industry interaction model presented in this book can be used as an effective mechanism for facilitating technology transfer. Industrial technology application centers may be established to serve as a unified point for linking technology sources with interested targets. The center will facilitate interactions between

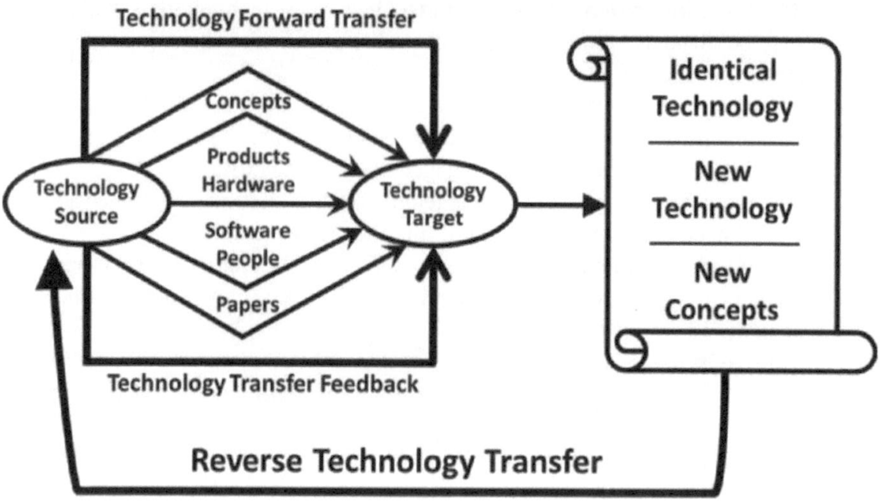

FIGURE 8.1 Technology transfer modes.

business establishments, academic institutions, and government agencies to identify important technological needs. With reference to Figure 8.1, technology can be transferred in one or a combination of the following strategies:

1. **Transfer of complete technological products:** In this case, a fully developed product is transferred from a source to a target. Very little product development effort is carried out at the receiving point. However, information about the operations of the product is fed back to the source so that necessary product enhancements can be pursued. The technology recipient generates product information which facilitates further improvement at the technology source. This is the easiest mode of technology transfer and the most tempting. Developing nations are particularly prone to this type of transfer. Care must be exercised to ensure that this type of technology transfer does not degenerate into "machine transfer." It should be recognized that machines alone do not constitute technology.
2. **Transfer of technology procedures and guidelines:** In this technology transfer mode, procedures (e.g., Blueprints) and guidelines are transferred from a source to a target. The technology blueprints are implemented locally to generate the desired services and products. The use of local raw materials and manpower is encouraged for local production. Under this mode, the implementation of transferred technology procedures can generate new operating procedures that can be fed back to enhance the original technology. With this symbiotic arrangement, a loop system is created whereby both the transferring and the receiving organizations derive useful benefits.
3. **Transfer of technology concepts, theories, and ideas:** This strategy involves the transfer of the basic concepts, theories, and ideas behind a given technology. The transferred elements can then be enhanced, modified,

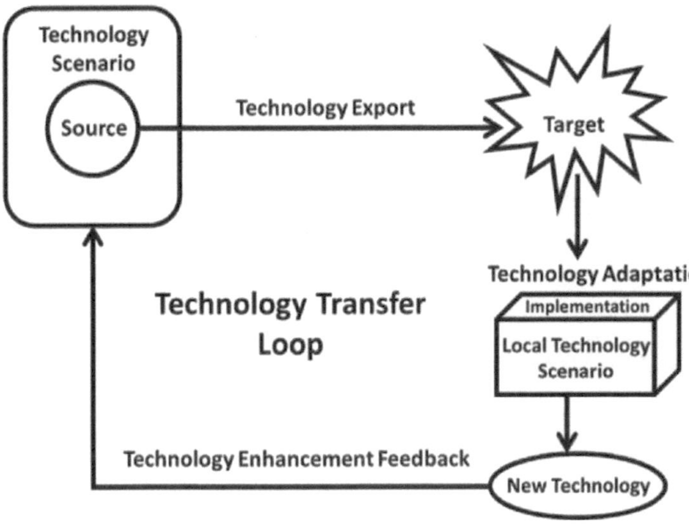

FIGURE 8.2 Local adaptation and enhancement of technology.

or customized within local constraints to generate new technological products. The local modifications and enhancements have the potential to generate an identical technology, a new related technology, or a new set of technological concepts, theories, and ideas. These derived products may then be transferred back to the original technology source as new technological enhancements. Figure 8.2 presents a specific cycle for local adaptation and modification of technology. An academic institution is a good potential source for the transfer of technology concepts, theories, and ideas.

It is very important to determine the mode in which technology will be transferred for manufacturing purposes. There must be a concerted effort by people to make the transferred technology work within local infrastructure and constraints. Local innovation, patriotism, dedication, and willingness to adapt technology will be required to make the technology transfer successful. It will be difficult for a nation to achieve industrial development through total dependence on transplanted technology. Local adaptation will always be necessary.

TECHNOLOGY CHANGE-OVER STRATEGIES

Any development project will require changing from one form of technology to another. The implementation of new technology to replace an existing (or a nonexistent) technology can be approached through one of several options. Some options are more suitable than others for certain types of technologies. The most commonly used technology change-over strategies include the following:

Parallel Change-over: In this case, existing technology and new technology operate concurrently until there is confidence that the new technology is satisfactory.

Direct Change-over: In this approach, the old technology is removed completely, and the new technology takes over. This method is recommended only when there is no existing technology or when both technologies cannot be kept operational due to incompatibility or cost considerations.

Phased Change-over: In this incremental change-over method, modules of the new technology are gradually introduced one at a time using either direct or parallel change-over.

Pilot Change-over: In this case, the new technology is fully implemented on a pilot basis in a selected department within the organization.

POST-IMPLEMENTATION EVALUATION

The new technology should be evaluated only after it has reached a steady-state performance level. This helps avoid the bias that may be present at the transient stage due to personnel anxiety, lack of experience, or resistance to change. The system should be evaluated for the following aspects:

- Sensitivity to data errors.
- Quality and Productivity.
- Utilization level.
- Response time.
- Effectiveness.

TECHNOLOGY SYSTEMS INTEGRATION

With the increasing shortages of resources, more emphasis should be placed on the sharing of resources. Technology resource sharing can involve physical equipment, facilities, technical information, ideas, and related items. The integration of technologies facilitates the sharing of resources. Technology integration is a major effort in technology adoption and implementation. Technology integration is required for proper product coordination. Integration facilitates the coordination of diverse technical and managerial efforts to enhance organizational functions, reduce costs, improve productivity, and increase the utilization of resources. Technology integration ensures that all performance goals are satisfied with a minimum expenditure of time and resources. It may require the adjustment of functions to permit the sharing of resources, development of new policies to accommodate product integration, or realignment of managerial responsibilities. It can affect both hardware and software components of an organization. Important factors in technology integration include the following:

- Unique characteristics of each component in integrated technologies.
- Relative priorities of each component in integrated technologies.
- How the components complement one another.
- Physical and data interfaces between the components.
- Internal and external factors that may influence integrated technologies.
- How the performance of the integrated system will be measured.

GOVERNMENT IN TECHNOLOGY TRANSFER

The malignant policies and operating characteristics of some of the governments in underdeveloped countries have contributed to the stunted growth of technology in those parts of the world. The governments in most developing countries control the industrial and public sectors of the economy. Either people work for the government or serve as agents or contractors for the government. The few industrial firms that are privately owned depend on government contracts to survive. Consequently, the nature of the government can directly determine the nature of industrial technological progress.

The operating characteristics of most governments perpetuate inefficiency, corruption, and bureaucratic bungles. This has led to a decline in labor and capital productivity in the industrial sectors. Using the Pareto distribution, it can be estimated that in most government-operated companies, there are eight administrative workers for every two production workers. This creates a non-productive environment that is skewed towards hyper-bureaucracy. The government of a nation pursuing industrial development must formulate and maintain an economic stabilization policy. The objective should be to minimize the sacrifice of economic growth in the short run while maximizing long-term economic growth. To support industrial technology transfer efforts, it is essential that a conducive national policy is developed.

More emphasis should be placed on industry diversification, training of the workforce, supporting financial structure for emerging firms, and implementing policies that encourage productivity in a competitive economic environment. Appropriate foreign exchange allocation, tax exemptions, bank loans for emerging businesses, and government-guaranteed low-interest loans for potential industrial entrepreneurs are favorable policies to spur growth and development in the industrial sector.

Improper trade and domestic policies have adversely affected industrialization in many countries. Excessive regulations that cause bottlenecks in industrial enterprises are not uncommon. The regulations can take the form of licensing, safety requirements, manufacturing value, added quota requirements, capital contribution by multinational firms, and high domestic production protection. Although regulations are necessary for industrial operations, excessive controls lead to low returns from the industrial sectors. For example, stringent regulations on foreign exchange allocation and control have led to the closure of industrial plants in some countries. The firms that cannot acquire essential raw materials, commodities, tools, equipment, and new technology from abroad due to foreign exchange restrictions are forced to close and lay off workers.

Price controls for commodities are used very often by developing countries, especially when inflation rates for essential items are high. The disadvantages involved in price control of industrial goods include restrictions on the free competitive power of available goods in relation to demand and supply, encouragement of inefficiency, promotion of dual markets, distortion of cost relationships, and increases in administrative costs involved in producing goods and services.

TEMPLATES FOR TECHNOLOGY TRANSFER

One way that a government can help facilitate industrial technology transfer involves the establishment of technology transfer centers within appropriate government agencies. A good example of this approach can be seen in the government-sponsored technology transfer program by the U.S. National Aeronautics and Space Administration (NASA). In the Space Act of 1958, the U.S. Congress charged NASA with a responsibility to provide the widest practical and appropriate dissemination of information concerning its activities and the results achieved from those activities. With this technology transfer responsibility, technology developed in the United States' space program is available for use by the nation's business and industry.

In order to accomplish technology transfer to industry, NASA established a Technology Utilization Program (TUP) in 1962. The technology utilization program uses several avenues to disseminate information on NASA technology. These avenues include the following:

- Complete, clear, and practical documentation is required for new technology developed by NASA and its contractors. These are available to the industry through several publications produced by NASA. An example is a monthly, Tech Briefs, which outlines technology innovations. This is a source of prompt technology information for industry.
- Industrial Application Centers (IAC) were developed to serve as repositories for vast computerized data on technical knowledge. The IACs are located at academic institutions around the country. All centers have access to a large database containing millions of NASA documents. With this database, the industry can have access to the latest technological information quickly. The funding for these centers is obtained through joint contributions from several sources including NASA, the sponsoring institutions, and state government subsidies. Thus, these centers can provide their services at very reasonable rates.
- NASA operates a Computer Software Management and Information Center (COSMIC) to disseminate computer programs developed through NASA projects. COSMIC, which is located at a university, has a library of thousands of computer programs. The center publishes an annual index of available software.

In addition to the specific mechanisms discussed above, NASA undertakes Application Engineering Projects. Through these projects, NASA collaborates with the industry to modify aerospace technology for use in industrial applications. To manage the application projects, NASA established a Technology Application Team (TAT), consisting of scientists and engineers from several disciplines. This team interacts with NASA field centers, industry, universities, and government agencies. The major mission of team interactions is to define important technology needs and identify possible solutions within NASA. NASA applications engineering projects are usually developed in a five-phase approach with go or no-go decisions made by NASA and the industry at the completion of each phase. The five phases are outlined below:

1. NASA and the Technology Applications Team meet with industry associations, manufacturers, university researchers, and public sector agencies to identify important technology problems that might be solved by aerospace technology.

2. After a problem is selected, it is documented and distributed to the Technology Utilization Officer at each of NASA's field centers. The officer in turn distributes the description of the problem to the appropriate scientists and engineers at the center. Potential solutions are forwarded to the team for review. The solutions are then screened by the problem originator to assess the chances for technical and commercial success.
3. The development of partnerships and a project plan to pursue the implementation of the proposed solution. NASA joins forces with private companies and other organizations to develop an applications engineering project. Industry participation is encouraged through a variety of mechanisms such as simple letters of agreement or joint endeavor contracts. The financial and technical responsibilities of each organization are specified and agreed upon.
4. At this point, NASA's primary role is to provide technical assistance to facilitate utilization of the technology. The costs for these projects are usually shared by NASA and the participating companies. The proprietary information provided by the companies and their rights to new discoveries are protected by NASA.
5. The final phase involves the commercialization of the product. With the success of commercialization, the project would have widespread impact. Usually, the final product development, field testing, and marketing are managed by private companies without further involvement from NASA.

Through this well-coordinated, government-sponsored technology transfer program, NASA has made significant contributions to the U.S. industry. The results of NASA's technology transfer abound in numerous consumer products, either in subtle forms or in clearly identifiable forms. Food preservation techniques constitute one area of NASA's technology transfer that has had a significant positive impact on society. Although the specific organization and operation of NASA technology transfer programs have changed in name or in deed over the years, the basic descriptions outlined above remain a viable template for how to facilitate manufacturing technology transfer. Other nations can learn from NASA's technology transfer approach. In a similar government-backed strategy, the U.S. Air Force Research Lab (AFRL) also has very structured programs for transferring non-classified technology to the industrial sector.

The major problem in developing nations is not the lack of good examples to follow. Rather, the problem involves not being able to successfully manage and sustain a program that has proven successful in other nations. It is believed that a project management approach can help in facilitating success with manufacturing technology transfer efforts.

BUILDING AN INDUSTRIAL BASE

Most developing nations depend on technologies transferred from developed nations to support their industrial base. This is partly due to a lack of local research and development programs, development funds, and workforce necessary to support such activity. Advanced technology is desired by most industries in developing countries because of its potential to increase output. The adaptability of advanced technology to industries in a developing country is a complex and difficult task. Evidence in most manufacturing firms that operate in developing countries reveal that advanced technology can lead to machine down time because local plants do not have the maintenance and repair facilities to support the use of advanced technology.

In some situations, most firms cannot afford the high cost of maintenance associated with the use of foreign technology. One way to solve the transfer of technology problem is by establishing local design centers for the developing nation's industrial sectors. These centers will design and adapt technology for local usage. In addition, such centers will also work on adapting fully assembled machinery from developed countries. However, the fertile ground for the introduction of appropriate technology is where people are already organized under a good system of government, production, marketing, and continuing improvement in standard of living. Developing countries must place more emphasis on the production of useful, consumable goods and services. One useful strategy to ensure a successful transfer of technology is providing training services that will ensure proper repair and maintenance of technology hardware. It is important that a nation trying to transfer technology should have access to a broad-based body of technical information and experience. A plan for technical information sharing between suppliers and users must be assured. The transfer of technology also requires a reliable liaison between the people who develop the ideas, their agents, and the people who originate the concepts. Technology transfer is only complete when the technology becomes generally accepted in the workplace. Local efforts are necessary in tailoring technological solutions to local problems. Technicians and engineers must be trained to assume the role of technology custodians so that implementation and maintenance problems are minimized. A strategy for minimizing the technology transfer disconnection is to set up central repair shops dedicated to making spare parts and repairing equipment on a timely basis to reduce industrial machine down time. If the utilization level of equipment is increased, there will be an increase in the productive capacity of the manufacturer. Improving maintenance and repair centers in developing countries will provide an effective way of assisting emerging firms in developing countries where dependence on transferred technology is prevalent. There should also be a strategy to develop appropriate local technology to support the goals of industrialization. This is important because fully transferred technology may not be fully suitable or compatible with local product specifications. For example, many nations have experienced the failure of transferred food processing technology because the technology was not responsive to the local diets, ingredients, and food preparation practices. One way to accomplish the development of local technology is to encourage joint research efforts between academic institutions and industrial firms. Some of the chapters in this book explicitly address university-industry collaborations. The design centers suggested earlier can help in this process. A chapter in this book presents a case example of the Industrial Development Center (IDCs) in Nigeria. In addition to developing new local technologies, existing technologies should be calibrated for local usage and the higher production level required for industrialization.

The government of developing nations must assume leadership roles in encouraging research and development activities by awarding research grants to universities and private organizations geared towards seeking better ways for developing and adapting technologies for local usage. Effective innovations and productivity improvement cannot happen without adequate public and private sector policies. A nation that does not have an effective policy for productivity management and technology advancement will always find itself in a cycle of unstable economy and business crisis. Increases in real product capital, income level, and quality of life are desirable goals that are achievable through effective policies that are executed properly. The following recommendations are offered to encourage industrial growth and technological progress:

1. Encourage a free enterprise system that believes in and practices fair competition. Discourage protectionism and remove barriers to allow free trade.
2. Avoid nationalization of the assets of companies jointly developed by citizens of developing countries and multinationals. Encourage joint industrial ventures among nations.
3. Both public and private sectors of the economy should encourage and invest in improving national education standards for citizens at various levels.
4. Refrain from dependence on borrowed money and subsidy programs. Create productive enterprises locally that provide essential commodities for local consumptions and exports.
5. Both public and private sectors should invest more in systems and programs as well as research and development that generates new breakthroughs in technology and methods for producing food rather than war instruments.
6. The public sector should establish science and technology centers to foster the development of new local technology, productivity management techniques, and production methodologies.
7. Encourage strong partnerships between government, industry, and academic communities in formulating and executing national development programs.
8. Governments and financial institutions should provide low-interest loans to entrepreneurs willing to take risk in producing essential goods and services through small-scale industries.
9. Implement a tax structure that is equitable and provides incentives for individuals and businesses that are working to expand employment opportunities and increase the final output of the national economy.
10. Refrain from government control of productive enterprises. Such controls only create grounds for fraud and corruption. Excessive regulations should be discouraged.
11. Periodically assess the ratio of administrative workers to production workers and administrative workers to service workers in both private and public sectors. Implement actions to reduce excessive administrative procedures and bureaucratic bottlenecks that impede productivity and technological progress.
12. Encourage organizations and firms to develop and implement strategies, methods, and techniques in a framework of competitive and long-term performance.
13. Trade policy laws and regulations should be developed and enforced in a framework that recognizes fair competition in a global economy.
14. Create a national productivity, science, and technology council to facilitate the implementation of good programs, enhance cooperation between

private and public sectors of the economy, redirect the economy towards growth strategies, and encourage education and training of the work force.

15. Implement actions that insure stable fiscal, monetary, and income policies. Refrain from wage and price control by political means. Let the elements of the free enterprise system control inflation rate, wages, and income distribution.
16. Encourage morale standards that take pride in excellence, work ethics, and value systems that encourage pride in consumer products produced locally.
17. Encourage individuals and businesses to protect full employment programs and maintain income levels by investing in local ventures rather than exporting capital abroad.
18. Both public and private sectors of the economy should encourage and invest in re-training the workforce as new technology and techniques are introduced for productive activities.
19. Make use of the expertise of nations that are professionally based abroad. This is an excellent source of expertise for local technology development.
20. Arrange for annual conferences, seminars, and workshops to exchange ideas between researchers, entrepreneurs, practitioners, and managers with focus on the processes required for industrial development.

PICKING THE RIGHT TECHNOLOGY

The question of which technology is appropriate to transfer in or transfer out is relevant for technology transfer considerations. While several methods of technology selection are available, this book recommends methods that combine qualitative and quantitative factors. The Analytical Hierarchy Process (AHP) is one such method. Another useful, but less publicized method is the PICK chart. The PICK chart was originally developed by Lockheed Martin to identify and prioritize improvement opportunities in the company's process improvement applications. This technique is just one of the several decision tools available in process improvement endeavors. It is a very effective technology selection tool used to categorize ideas and opportunities. The purpose is to qualitatively help identify the most useful ideas. A 2x2 grid is normally drawn on a white board or large flip chart. Ideas that were written on sticky notes by team members are placed on the grid based on a group assessment of the payoff relative to the level of difficulty. The PICK acronym comes from the labels for each of the quadrants of the grid: **P**ossible (easy, low payoff), **I**mplement (easy, high payoff), **C**hallenge (hard, high payoff), and **K**ill (hard, low payoff). The PICK chart quadrants are summarized as follows:

Possible (easy, low payoff) → Third quadrant
Implement (easy, high payoff) → Second quadrant
Challenge (hard, high payoff) → First quadrant
Kill (hard, low payoff). → Fourth quadrant

The primary purpose is to help identify the most useful ideas, especially those that can be accomplished immediately with little difficulty. These are called "Just-Do-Its." The general layout of the PICK chart grid is shown in Figure 8.3. The PICK process is normally done subjectively by a team of decision makers under a group decision

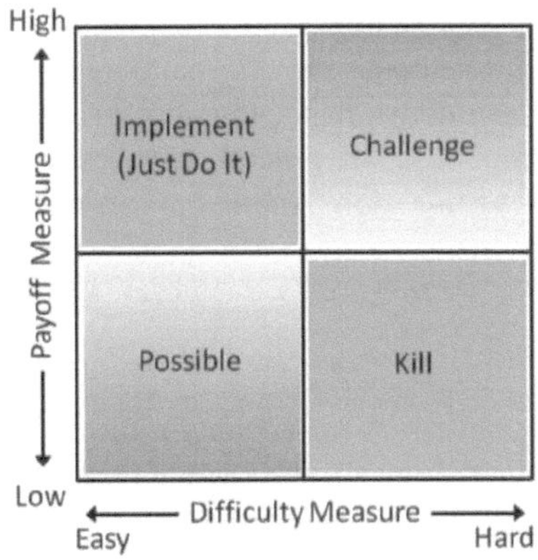

FIGURE 8.3 Basic layout of the PICK chart.

process. This can lead to bias and a protracted debate of where each item belongs. It is desired to improve the efficacy of the process by introducing some quantitative analysis. Badiru and Thomas (2013) present a methodology to achieve a quantification of the PICK selection process. The PICK chart is often criticized for its subjective rankings and lack of quantitative analysis. The approach presented by Badiru and Thomas (2013) alleviates such concerns by normalizing and quantifying the process of integrating subjective rankings by those involved in the group PICK process. Human decision is inherently subjective. All we can do is develop techniques to mollify the subjective inputs rather than compounding them with subjective summarization.

The placement of items into one of the four categories in a PICK chart is done through expert ratings, which are often subjective and non-quantitative. To put some quantitative basis to the PICK chart analysis, Badiru and Thomas (2013) present the methodology of dual numeric scaling on the impact and difficulty axes. Suppose each technology is ranked on a scale of one to ten and plotted accordingly on the PICK chart. Then, each project can be evaluated on a binomial pairing of the respective rating on each scale. Note that a high rating along the x axes is desirable while a high rating along the y axis is not desirable. Thus, a composite rating involving x and y must account for the adverse effect of high values of y. A simple approach is to define $y' = (11\text{-}y)$, which is then used in the composite evaluation. If there are more factors involved in the overall project selection scenario, the other factors can take on their own lettered labeling (e.g., a, b, c, z, etc.). Then, each project will have an n-factor assessment vector. In its simplest form, this approach will generate a rating such as the following:

$$PICK_{R,i}(x, y') = x + y'$$

Where:

$PICK_{R,i}$(x,y) = PICK rating of project i (i = 1, 2, 3, …. , n).
n = number of projects under consideration.
x = rating along the impact axis ($1 \leq x \leq 10$).
y = rating along the difficulty axis $1 \leq y \leq 10$).
$y' = (11-y)$.

If $x + y'$ is the evaluative basis, then each technology's composite rating will range from two to twenty, with two being the minimum and twenty being the maximum possible. If *(x)(y)* is the evaluative basis, then each project's composite rating will range from one to one hundred. In general, any desired functional form may be adopted for the composite evaluation. Another possible functional form is:

$$\begin{aligned} PICK_{R,i}(x, y'') &= f(x, y'') \\ &= (x + y'')^2, \end{aligned}$$

where y'' is defined as necessary to account for the converse impact of the axes of difficulty. The above methodology provides a quantitative measure for translating the entries of a conventional PICK chart into an analytical technique to rank the technology alternatives, thereby reducing the level of subjectivity in the final decision. The methodology can be extended to cover cases where technology has the potential to create negative impacts, which may impede organizational advancement.

The quantification approach facilitates a more rigorous analytical technique compared to traditional subjective approaches. One concern is that although quantifying the placement of alternatives on the PICK chart may improve the granularity of relative locations on the chart, it still does not eliminate the subjectivity of how the alternatives are assigned to quadrants in the first place. This is a recognized feature of many decision tools. This can be mitigated using additional techniques that aid decision makers to refine their choices. The analytic hierarchy process (AHP) could be useful for this purpose. Quantifying subjectivity is a continuing challenge in decision analysis. The PICK chart quantification methodology offers an improvement over the conventional approach.

Although the PICK chart has been used extensively in industry, there are few published examples in open literature. The quantification approach presented by Badiru and Thomas (2013) may expand interest and applications of the PICK chart among technology researchers and practitioners. The steps for implementing a PICK Chart are summarized below:

Step 1: On a chart, place the subject question. The question needs to be asked and answered by the team at different stages to be sure that the data that is collected is relevant.

Step 2: Put each component of the data on a different note like a post-it or small cards. These notes should be arranged on the left side of the chart.

Step 3: Each team member must read all notes individually and consider their importance. The team member should decide whether the element should or should not remain a fraction of the significant sample. The notes are then removed and moved to the other side of the chart. Now, the data is condensed enough to be processed for a particular purpose by means of tools that allow groups to reach a consensus on priorities of subjective and qualitative data.

Step 4: Apply the quantification methodology presented above to normalize the qualitative inputs of the team.

Technology transfer is a great avenue to advancing industrialization (Badiru, 2012, 2016). This chapter has presented a variety of principles, tools, techniques, and strategies useful for managing technology transfer. Of particular emphasis in this chapter is the management aspects of technology transfer. The technical characteristics of the technology of interest are often well understood. What is often lacking is an appreciation of the technology management requirements for achieving a successful technology transfer. This chapter presents the management aspects of manufacturing technology transfer.

REFERENCES

Badiru, A. B. (2012), "Application of the DEJI Model for Aerospace Product Integration," ***Journal of Aviation and Aerospace Perspectives (JAAP)***, Vol. 2, No. 2, pp. 20–34, Fall 2012.

Badiru, A. B. (2016), **Global Manufacturing Technology Transfer: Africa-USA Strategies, Adaptations, and Management**, Taylor & Francis CRC Press, Boca Raton, FL.

Badiru, A. B., and M. Thomas (2013), "Quantification of the PICK Chart for Process Improvement Decisions," ***Journal of Enterprise Transformation***, Vol. 3, No. 1, pp. 1–15.

9 Workforce Framework for Re-Industrialization of Dayton Aviation

DAYTON'S AVIATION LEGACY

> The Wright Brothers created the single greatest cultural force since the invention of writing. The airplane became the first World Wide Web, bringing people, languages, ideas, and values together.
>
> **Bill Gates**

Dayton is not only the birthplace of aviation, but also a national hub in aerospace technology. Leonardo da Vinci (April 15, 1452 to May 2, 1519), a fifteenth-century Italian polymath (painter, draughtsman, engineer, scientist, theorist, sculptor, and architect), theorized that man could fly. It was over 400 years before the Wright Brothers of Dayton, Ohio, actually proved this theory in 1903. Thus, emerged the aviation legacy of Dayton. In its heyday, at the height of the innovation renaissance, Dayton was the place to be, by any world standard. The most technical and astute workforce was developed in Dayton, leading to an innumerable number of inventions that changed the entire world. After initial accolades of Dayton, over the years, Dayton's worldwide visibility and recognition waned, perhaps due to the ensuing lackadaisical and neglectful promotion of Dayton around the world. The younger generation, unless re-informed, may not realize the importance of Dayton in world workforce affairs. The desire advocated in this chapter is to embark on a contemporary workforce development for the re-industrialization of Dayton.

Many product developments and advancements came out of Dayton in the wake of the Wright Brothers' aviation breakthrough (Donais, 2002; Chollet, 2005; Badiru, 2024).

Ms. Jessica Short, Director of Dayton Regional STEM Center, spoke at a 2024 event at the Engineers Club of Dayton and conducted an activity to emphasize Dayton's aviation history. She asked the entire audience to stand up. She then proceeded to ask groups of the audience to sit down based on categories of technology usage:

"If you bought or opened a can of soda today, please sit down."
"If you used ice cubes made in a refrigerator's ice-cube tray, please sit down."
"If you fly into town from somewhere (anywhere) around the world, please sit down."
".. …please sit down."
"… please sit down."

DOI: 10.1201/9781003537755-9

On and on.

"If you rode in any type of vehicle to get to this building today, please sit down."

By this time, everyone in the audience had taken his or her seat.
The speaker then said:

"That is Dayton!"
"All the items you just acknowledged to have benefited from all that originated right here in Dayton."

Such is the widespread legacy of Dayton's innovation. It was a brilliant way of conveying the message of what Dayton means to society, then and now.

ORIGIN OF DAYTON

Following the 1795 peace treaty with the Shawnee Indians, signed in Fort Greenville (now Greenville, Ohio), the area was opened to white settlement. The town was laid out by a group of Revolutionary War veterans, including Jonathan Dayton from New Jersey, for whom it was named. It developed as a river port for the shipment of agricultural produce, mainly to New Orleans. The opening of the Miami and Erie Canal, from Dayton to Cincinnati in 1829, and the arrival of a railroad to Springfield in 1851 stimulated Dayton's commercial and industrial growth. The town became the home of the cash register after the mechanical money drawer was invented there in 1879 by James Ritty and perfected by John Henry Patterson in the 1880s. In addition, the automobile self-starter was developed there by Charles F. Kettering, who, along with Edward A. Deeds, also produced ignition systems and electric lighting equipment for farms. In 1892, Wilbur and Orville Wright opened their bicycle repair shop in Dayton, where they conducted experiments that led to the first sustained and controlled flight of a powered airplane, at Kitty Hawk, North Carolina, in 1903. A monolith has been erected in memory of the brothers, who are buried in the city's Woodland Cemetery.

DAYTON PEACE ACCORD

One gem of a topic that is rarely promoted within and outside the "gem city" is the positive role that Dayton played in the Peace Accord. The General Framework Agreement for Peace in Bosnia and Herzegovina, also known as the Dayton Agreement or the Dayton Accords, and colloquially known as the Dayton is the peace agreement reached at Wright-Patterson Air Force Base near Dayton, Ohio (where all three authors of this book once worked). The treaty was finalized on November 21, 1995 and formally signed in Paris, on December 14, 1995 (Goodby, 1996; Woodward, 1996; Caplan, 2000; Curran et al, 2004; Tuathail et al, 2006; Parish, 2007; Belloni, 2009; Chivvis, 2010; Daalder 2014). These accords put an end to the three-and-a-half-year-long Bosnian War, which was part of the much larger Yugoslav Wars.

The warring parties agreed to peace and a single sovereign state, known as Bosnia and Herzegovina, composed of two parts, the largely Serb-populated Republika Srpska and mainly Croat-Bosniak-populated Federation of Bosnia and Herzegovina.

This agreement has been criticized for creating ineffective and unwieldy political structures and entrenching the ethnic cleansing of the previous war. This negative aspect may be a subtle reason why many don't mention the Dayton Peace Accord openly in many quarters. Regardless of this, there is no denying the role, visibility, recognition, and legacy that Dayton achieved due to the protracted discussions and negotiations. The fact is that Dayton and a select portion of the workforce did play a visible role in this international affair. Without a local workforce (blue collar and white collar) to support the efforts, this resolution might not have happened in Dayton.

Though basic elements of the Dayton Agreement were proposed in international talks as early as 1992, these negotiations were initiated following the unsuccessful previous peace efforts and arrangements of the August 1995 Croatian military Operation Storm and its aftermath, as well as the government military offensive against the Republika Srpska, conducted in parallel with NATO's Operation Deliberate Force. During September and October 1995, world powers (especially the United States and Russia) gathered in the Contact Group and pressured the leaders of the three sides to attend settlement negotiations. Dayton, Ohio, was eventually chosen as the venue.

Discussion began with an outline of key points presented by the U.S. in a team led by National Security Adviser Anthony Lake in visits to London, Bonn, Paris, and other European stops 10–14 August 1995. These included Sochi, to consult the Russian Foreign Minister Andrei Kozyrev. Lake's team handed off to a separate U.S. inter-agency group led by Assistant Secretary of State Richard Holbrooke, who went on to negotiate with Balkan leaders in their capitals. The Holbrooke crew conducted five rounds of intense shuttle diplomacy from August to October, including short conferences in Geneva and New York that resulted in the parties' adoption of principles for a settlement on 8 and 26 September, respectively.

The Dayton conference took place from 1st–21st November 1995. The main participants from the region were the President of the Republic of Serbia, Slobodan Milošević (whom the Bosnian Serbs had previously empowered to represent their interests), the President of Croatia, Franjo Tuđman, and the President of Bosnia and Herzegovina, Alija Izetbegović, with his Foreign Minister Muhamed Šaćirbeg.

The peace conference was led by U.S. Secretary of State Warren Christopher and negotiator Richard Holbrooke, with two co-chairmen in the form of EU Special Representative Carl Bildt and the First Deputy Foreign Minister of Russia, Igor Ivanov. A key participant in the U.S. delegation was General Wesley Clark. The head of the UK's team was Pauline Neville-Jones, political director of the Foreign and Commonwealth Office. The UK military representative was Col. Arundell David Leakey. Paul Williams, through the Public International Law & Policy Group (PILPG), served as legal counsel to the Bosnian Government delegation during the negotiations. Holbrooke spoke of the "immense difficulty of engaging the Bosnian government in a serious negotiation."

The secure site was chosen to remove all the parties from their comfort zone, without which they would have little incentive to negotiate or reduce their ability to negotiate through the media and securely house over 800 staff and attendants. Curbing the participants' ability to negotiate via the media was a particularly important consideration. Holbrooke wanted to prevent posturing through early leaks to the press.

After having been initialed in Dayton, Ohio, on November 21, 1995, the full and formal agreement was signed in Paris on December 14, 1995 and witnessed by Spanish Prime Minister Felipe González, French President Jacques Chirac, U.S. President Bill Clinton, UK Prime Minister John Major, German Chancellor Helmut Kohl, and Russian Prime Minister Viktor Chernomyrdin.

The agreement's main purpose is to promote peace and stability in Bosnia and Herzegovina, and to endorse regional balance in and around the former Yugoslavia.

The present political divisions of Bosnia and Herzegovina and its structure of government were agreed upon. A key component of this was the delineation of the Inter-Entity Boundary Line.

The State of Bosnia and Herzegovina is composed of the Federation of Bosnia-Herzegovina and of the Republika Srpska. Bosnia and Herzegovina is a complete state, as opposed to a confederation. No entity or entities could ever be separated from Bosnia and Herzegovina unless by due legal process. Although highly decentralized in its entities, it would still retain a central government, with a rotating State Presidency, a central bank, and a constitutional court.

The agreement mandated a wide range of international organizations to monitor, oversee, and implement components of this agreement. The NATO-led IFOR (Implementation Force) was responsible for implementing the military aspects of the agreement and deployed on December 20, 1995, taking over the forces of the UNPROFOR. The Office of the High Representative was charged with the task of civil implementation. The Organization for Security and Co-operation in Europe was charged with organizing the first free elections in 1996.

On October 13, 1997, the Croatian 1861 Law Party and the Bosnia-Herzegovina 1861 Law Party requested the Constitutional Court of Bosnia and Herzegovina to annul several decisions and to confirm one decision of the Supreme Court of the Republic of Bosnia and Herzegovina. More importantly, to review the constitutionality of the General Framework Agreement for Peace in Bosnia and Herzegovina, since it was alleged that the agreement violated the Constitution of Bosnia and Herzegovina in a way that it undermined the integrity of the state, and could cause the dissolution of Bosnia and Herzegovina. The Court reached the conclusion that it is not competent to decide the dispute regarding the mentioned decisions since the applicants were not subjects that were identified in Article VI.3 (a) of the Constitution on those who can refer disputes to the Court. The Court also rejected the other request. The Constitutional Court is not competent to evaluate the constitutionality of the General Framework Agreement, as the Constitutional Court has in fact been established under the Constitution of Bosnia and Herzegovina to uphold this Constitution. The Constitution of Bosnia and Herzegovina was adopted as an annex to the General Framework Agreement for Peace in Bosnia and Herzegovina, and consequently there cannot be a conflict or a possibility for controversy between this Agreement and the Constitution of Bosnia and Herzegovina.

It was one of the earliest cases in which the Court had to deal with the question of the legal nature of the Constitution. By making the remark in the manner of obiter dictum concerning the Constitution and the rest of the peace agreement, the Court actually "established the ground for legal unity" of the entire peace agreement, which further implied that all the annexes are in the hierarchical equality. In later decisions, the Court confirmed that by using other annexes of the peace agreement as a direct base for the analysis, not only in the context of systematic interpretation of Annex IV. However, since the Court rejected the presented request of the appellants, it did not go into details concerning the controversial questions of the legality of the process in which the new Constitution (Annex IV) came to power and replaced the former Constitution of the Republic of Bosnia and Herzegovina. The Court used the same reasoning to dismiss a similar claim in a later case.

The immediate purpose of the agreement was to freeze the military confrontation and prevent it from resuming. It was therefore defined as a "construction of necessity."

The Dayton Agreement aimed at allowing Bosnia and Herzegovina to move from an early post-conflict phase through reconstruction and consolidation, adopting a consociational power-sharing approach. Scholars such as Canadian professor Charles-Philippe David call Dayton "the most impressive example of conflict resolution." American scholar Howard M. Hensel states that "Dayton represents an example of a conflict resolution negotiation that was successful." However, Patrice C. McMahon and Jon Western write that "As successful as Dayton was at ending the violence, it also sowed the seeds of instability by creating a decentralized political system that undermined the state's authority."

High Representative Wolfgang Petritsch argued in 2006 that the Dayton framework has allowed the international community to move "from state building via institutions and capacity building to identity building," putting Bosnia and Herzegovina "on the road to Brussels."

The Dayton Agreement has been the subject of criticism since its inception, including:

A complicated government system – as part of the Dayton agreement, Bosnia was divided regionally between two "entities" within a consociational democracy, which was established to ensure the political representation and power of all sides. This can lead to an unproductive government where every important issue is deadlocked within the central government as each party is championing opposing priorities that are based on ethnic policies and not shared ideals.

Dependency and control of international actors – Dayton was very much an international vision, led by the United States who supported an end to the war, but did not allow Bosnian leaders to negotiate an ending to the war, therefore leaving no incentive in the afterward peacebuilding process and no area for leaders to discuss the underlying root causes of the conflict. International actors also played an extensive role in shaping the postwar agenda in Bosnia. The international community invests millions of dollars in BiH yearly through NGOs. However, this stifles the impact of local actors and the development of civil society. Instead, the international community should invest in local actors, youth activists, and democratization projects. The influx of NGOs and international actors to kick start investment in the country post-war also failed to kick start the economy, with Bosnia suffering from poor economic growth (2% in 2015). The lack of economic development has been attributed to poor coordination between international actors and lack of consideration for local capacity.

Ending the war, but not promoting peace – the primary aim of Dayton was to stop the war, but the agreement was only meant to be a temporary measure while a long-term plan was developed. The Dayton Agreement was the thirty-fifth attempt at a ceasefire following thirty-four other failed attempts. While Dayton halted the conflict and there has not been a resurgence of violence, the stability in the conflict does not give an accurate assessment of peace. There is negative peace in BiH, meaning there is no open conflict or violence. However, there is no positive peace, as conditions that eliminate the causes of violence have not been reached. There is still currently an international military presence, EUFOR Althea, responsible for overseeing compliance with aspects of the Dayton Agreement. The Dayton Agreement provided peace by re-establishing and codifying division. Enforcing such peace can be seen as highlighting the still deep-rooted tensions in the country, with Dayton covering the cracks of a fractured society that could be plunged back into conflict as soon as military forces left.

Consociational Democracy – The Dayton Agreement established a consociational democracy in Bosnia and Herzegovina. This means that each group is ensured representation and power. This incentivized the end of the Bosnian War, but first requires collaboration or reconciliation for the government to function. Bosnia and Herzegovina operates with a three-member president role. There is a Croat, Bosniak, and Serb president. Similar quotas and rules apply to the two legislative bodies.

Entrenching territorialized ethnicity – The Agreement was underpinned by a territorialized definition of ethnicity that divided Bosnia and Herzegovina into three constitutive nations and two distinct entities based on ethno-nationalist identities.

According to survey results from a 2020 study, "in each of the three main ethnic groups of Bosnia, more people would have voted for Dayton than against it."

On February 13, 2008, the head of the Presidency of Bosnia-Herzegovina, Željko Komšić, said that the original Dayton Agreement was lost from the Presidency's archive. High Representative for Bosnia-Herzegovina Miroslav Lajčák said, "I don't know whether the news is sad or funny." On November 16, 2009, the French Foreign Ministry delivered the certified copy of the Dayton agreement to the French embassy in Sarajevo. The copy was later transferred to the Ministry of Foreign Affairs of Bosnia-Herzegovina. The stolen original was found in 2017 in a private residence in Pale, resulting in the arrest of the person who was trying to sell it.

2025 SPRING NATO ASSEMBLY IN DAYTON

To commemorate the thirtieth anniversary of the Dayton Peace Accord, the Spring 2025 NATO Parliamentary Assembly took place in Dayton, Ohio. The bragging point for the NATO assembly is that the Dayton region is home to Wright-Patterson Air Force Base, where the Dayton Accords, ending the Bosnian War, were signed in 1995. The NATO Parliamentary Assembly was created in 1955 to serve as a link between the North Atlantic Treaty Organization (NATO) and the parliaments of the now thirty-two NATO nations, with the most recent additions being Finland (2023) and Sweden (2024). The United States has not hosted a session since the Annual Session in Orlando, FL, in the Fall of 2003. The U.S. has never hosted a Spring Session of the Assembly until Dayton 2025, commemorating the thirtieth anniversary of the Dayton Peace Accords. Mike Turner, U.S. Congressman from Ohio, says it best:

> Israel and Gaza, and all of the national security threats and international street threats that you see around the world will be discussed and deliberated right here in Dayton, Ohio, next Memorial Day.
>
> **Mike Turner, U.S. Congressman from Ohio**

This chalks it up for the local Dayton workforce. More workforce development is in order to maintain Dayton's aviation and technological legacy.

A WORKFORCE DEVELOPMENT PLAN

Nicole Schreiber-Shearer (2024) presents a practical template for a workforce development plan.

By Nicole Schreiber-Shearer, Future of Work Specialist at GloatFirst Published – September 11, 2024 | Last Updated – September 12, 2024

In a world as ever-changing as ours, some leaders may wonder whether creating a plan is even worth it. If everything is going to change anyway, what's the point in laying out a map of what to do next?

In actuality, planning is more important now than it's ever been before. Change may be our only constant, but that doesn't mean we can't create—and don't need—plans to navigate these changes effectively.

In fact, with the right workforce development plan, businesses will be equipped with the tools and infrastructure necessary to capitalize on emerging opportunities and conquer any curveballs that come their way. Since not all approaches are created equally, we'll break down what a top-performing workforce development strategy looks like and share some tips so that you can level up your own approach.

What Is a Workforce Development Plan?

A workforce development plan is a formal or informal program implemented by businesses to support career development. Traditionally, these plans focused on building specific skills, usually through training programs. Today, common strategies for employee development have expanded, with the most comprehensive workforce development plans now including educational opportunities, mentorships, coaching programs, forums, and webinars for employees to share best practices.

Some workforce development plans also build partnerships with other organizations to create opportunities for employees to gain new skills and certifications, often through shadowing, mentoring, and apprenticeship programs.

What Is the Purpose of a Workforce Development Plan?

A workforce development plan is important because it creates a single source of truth that dictates training and development goals, as well as the steps leaders need to take to achieve these objectives. Workforce development plans keep leaders across the organization aligned on what they're trying to accomplish and the skills their people will need to turn big-picture goals into tangible realities.

Without a workforce development plan, employees may complete training that doesn't align with organizational priorities or the direction of the business. Even worse, they may not pursue any development opportunities whatsoever, contributing to ever-widening knowledge gaps that will eventually slow output levels and revenue growth to a halt.

What Should Be Included in an Effective Work Development Plan?

While every work development plan will differ depending on business priorities and the needs of the organization, most are comprised of the following components:

#1. Training and Upskilling

Upskilling and training describe intentional learning processes where employees gain exposure to deeper knowledge, typically through content-based coursework and experiential learning opportunities. They enable employees to develop their mastery of both technical and soft skills. The goal of these initiatives is to help people move into more advanced roles and ensure workforces have the skills their businesses need to accomplish strategic priorities.

#2. Career Development Pathways

When employees can see the various directions their careers can go in, they will be inspired to continue developing their skills to turn their professional goals into reality. The best career pathing tools, found within an agile workforce OS, can pinpoint the exact skills people need to hone so they can step into their dream roles, as well as projects and gigs within their company that will help them advance their careers.

#3. Succession Planning

Part of a successful workforce development plan is thinking ahead and determining what training employees will need to eventually step into high-priority leadership roles. Succession planning ensures that productivity and employee morale don't take a hit during periods of management turnover, because there will always be employees with the skills and experience required to lead with confidence.

#4. Performance Management

HR leaders must take stock of employees' progress and monitor how the work they're doing aligns with expectations for their roles. Managers should meet with their direct reports regularly to discuss how various projects and deliverables are going and identify any missing skills that are holding people back. By routinely evaluating employees' performance, leaders can adjust their workforce development plan to ensure all-star workers are on track to move into high-priority roles and employees whose performance needs improvement are getting the help they need.

5 Steps to Creating an Effective Workforce Development Plan

Looking to level up your workforce development plan? Consider the following best practices:

#1. Assess Current Capabilities

Before you can create a workforce development plan that addresses existing and emerging skill gaps, you need to know where your people's skills lie to begin with. Traditionally, gaining a bird's-eye view of workforce capabilities has been hard to come by because skills information is often siloed into a few different HR systems. Most taxonomies are also out of date, leading to skill-building strategies that may not address the actual needs of the organization.

Rather than settling for this subpar understanding of workforce skills, companies are harnessing AI-powered systems like Gloat's Skills Foundation to gain visibility into all their people's capabilities. These platforms are updated in real-time and pull from an employee's LinkedIn profile or CV, in turn capturing the full range of knowledge and experience they're bringing to the table.

#2. Forecast Future Workforce Needs

Leaders must keep a pulse on which roles in their organizations are becoming outdated, as well as new "sunrising" positions that are becoming more popular. With help from AI-powered skills intelligence systems, executives can gain insight into the capabilities that are surging in demand and the jobs that will be needed in the years to come. Armed with these insights, leaders can then create development pathways that will enable employees—particularly those in roles that are becoming less critical—to build the expertise necessary to move into these sought-after sunrising roles.

#3. Set Clear Objectives and Goals

Following an analysis of workforce skills, leaders must set goals that will guide their development plan. Skills benchmarking is one of the most useful tools for this goal-setting process because it allows leaders to measure employees' skills to see how they compare to organizational and industry-wide standards.

This analysis can help HR and L&D leaders determine whether an employee has the skill sets required to perform a role well and identify upskilling opportunities to help bridge any existing gaps. Skill benchmarking is a relatively new phenomenon made possible by the rise of skills intelligence systems. As these platforms go mainstream, more organizations can evaluate their workforce's capabilities against the competencies their competitors possess and adjust their strategies accordingly.

#4. Develop Strategies and Action Plans

Once leaders have an in-depth understanding of their workforce's skills, they must architect a plan to help employees build the competencies outlined in their development strategies. Employees need multiple pathways to hone new skills, including having the opportunity to pitch in on projects and gigs so they can gain on-the-job experience.

A talent marketplace that matches people to open opportunities based on the experiences they have and the skills they wish to build can be a game-changer for employee development. The platforms can also help employees find mentors in their organizations so they can receive one-on-one guidance as they expand their skill sets.

#5. Establish Timelines and Milestones

A workforce development plan isn't complete once leaders have drafted it. Instead, executives must carefully monitor their organization's skill-building efforts to determine who's participating in development opportunities and how these opportunities are bringing employees closer to reaching their overarching skill goals.

Leaders can use skills intelligence systems to get an idea of which capabilities are growing more important, and which are declining according to market and industry trends. Based on this information, leaders may choose to deprioritize developing a certain competency, so they can shift gears and focus on skills that are more in demand.

The Role of Leadership in Workforce Development

The quality of your leadership team can make or break the success of your workforce development plan. If leaders fail to engage their workforce and build enthusiasm for learning and skill-building, the development plan will likely fall flat.

In contrast, leaders who communicate transparently and emphasize how skill-building will not only help the business but also enable employees to reach their career goals, will motivate their workforce to pursue their plan. Leaders should give employees ample opportunities to ask questions about the workforce development strategy and to share feedback to help guide the initiative.

The Long-Term Impact of Effective Workforce Development

All too often, leaders don't realize they have a skills shortage until it's too late. Rather than letting skill gaps slow productivity and diminish performance, savvy executives can start executing workforce development plans to tackle any shortages head-on.

Simultaneously, creating a workforce skills plan sends a clear signal to your entire organization that leaders view continuous learning and development as a top priority. When employees are aligned on the skills their organizations need to be successful—both now and in the future—they will be inspired to continue training so they can play a part in helping the business reach its overarching objectives. A workforce development strategy creates a sense of transparency and reminds employees that they all must come together for this common goal.

Want to learn how one leading pharmaceutical company took workforce development to the next level? Listen to this conversation with Novartis's Global Head of Talent Markus Graf and The Josh Bersin Company.

A RAY OF ADVANCEMENT

One re-industrialization pathway that has morphed into a ray to actualization is the 2025 plan by Joby Aviation to expand its manufacturing activities in Dayton, along with the existing expansion in California. Joby reaffirmed its plan to deliver its first parts from a Dayton International Airport facility by mid-2025. This is befitting because Dayton was where Aviation research, development, manufacturing, and exporting all started at the turn of the Century. The Santa Cruz, California-based Joby Aviation is pumping close to $1 billion into its manufacturing pursuits. It expects to deliver its first air taxi passengers in 2026. Dayton is primed to tap into that economic largesse. It is expected that the city and its progressively-minded leaders will capitalize on this new opportunity for re-industrialization of Dayton.

REFERENCES

Badiru, A. B. (June 2024), "CHIP on Our Shoulder Requires Strategy for Workforce Development," ***ISE Magazine***, p. 24.

Belloni, R. (2009), "Bosnia: Dayton Is Dead! Long Live Dayton!" ***Nationalism and Ethnic Politics***, Vol. 15, No. 3–4, pp. 355–375.

Caplan, R. (2000), "Assessing the Dayton Accord: The Structural Weaknesses of the General Framework Agreement for Peace in Bosnia and Herzegovina," ***Diplomacy and Statecraft***, Vol. 11, No. 2, pp. 213–232.

Chivvis, C. S. (2010), "The Dayton Dilemma," ***Survival***, Vol. 52, No. 5, pp. 47–74; doi:10.1080/00396338.2010.522096.

Chollet, D. (2005), **The Road to the Dayton Accords**, Palgrave Macmillan, New York, NY.

Curran, D., J. K. Sebenius, and M. Watkins (2004), "Two Paths to Peace: Contrasting George Mitchell in Northern Ireland With Richard Holbrooke in Bosnia–Herzegovina," ***Negotiation Journal***, Vol. 20, No. 4, pp. 513–537.

Daalder, I. H. (2014), **Getting to Dayton: the Making of America's Bosnia Policy**, Brookings Institution Press, New York, NY.

Donais, T. (2002), "The Politics of Privatization in Post-Dayton Bosnia," ***Southeast European Politics***, Vol. 3, No. 1, pp. 3–19.

Goodby, J. E. (1996), "When War Won Out: Bosnian Peace Plans Before Dayton," ***International Negotiation***, Vol. 1, No. 3, pp. 501–523.

Parish, M. (2007), "The Demise of the Dayton Protectorate. Inside the Bosnian Crisis: Documents and Analysis," ***Journal of Intervention and Statebuilding***, Vol. 1, pp. 11–23.

Schreiber-Shearer, N. (2024), Building a Future-Ready Team: Effective Workforce Development Strategies, Blog Posting, Accessed 22 November 2024. https://gloat.com/blog/workforce-development-plan/

Tuathail, G. Ó., J. O'Loughlin, and D. Djipa (2006), "Bosnia-Herzegovina Ten Years after Dayton: Constitutional Change and Public Opinion," ***Eurasian Geography and Economics***, Vol. 47, No. 1, pp. 61–75; doi:10.2747/1538-7216.47.1.61.

Woodward, S. L. (1996), **Implementing Peace in Bosnia and Herzegovina: a Post-Dayton Primer and Memorandum of Warning**, Foreign Policy Studies Program, Brookings Institution, New York, NY.

10 An Integrated Systems Model for Workforce Development

As markets grow and technology advances, workforce development must move beyond traditional approaches to work design and process improvements to achieve or maintain global leadership.

– Sharon Bommer

TECHNOLOGY, EFFICIENCY, AND QUALITY

Systems quality is at the intersection of efficiency, effectiveness, and productivity. Efficiency provides the framework for quality in terms of resources and inputs required to achieve the desired level of quality. Effectiveness comes into play with respect to the application of product quality to meet specific needs and requirements of an organization. Productivity is an essential factor in the pursuit of quality as it relates to the throughput of a production system. To achieve the desired levels of quality, efficiency, effectiveness, and productivity, a new research framework must be adopted. In this chapter, we present a quality enhancement model for quality design, evaluation, justification, and integration (DEJI) based on a product development application. This model is relevant for research efforts in quality engineering and technology applications and other systems engineering applications.

The DEJI model is a systems engineering framework for work design that can be used to incorporate cognitive ergonomics, as defined in Chapter 5, into modern manufacturing processes. As technological advancements continue to reshape manufacturing, particularly with the transition of additive manufacturing (AM) from prototyping to mass production, there is an increasing need to prioritize cognitive ergonomics in system design.

Additive manufacturing's shift from a prototyping tool to a platform for rapid manufacturing is redefining the industry landscape, presenting new opportunities and challenges. This transformation requires a reimagining of work design to better support human operators. By integrating cognitive ergonomics principles and leveraging modern technologies like Artificial Intelligence, manufacturing processes can be optimized to enhance the following:

- Quality.
- Speed.
- Safety.
- Usability.

DOI: 10.1201/9781003537755-10

The DEJI model provides a systematic framework for evaluating and integrating cognitive ergonomic factors into work system design. This approach ensures that human capabilities and limitations are central to decision-making, enabling the development of systems that not only maximize productivity but also support operator well-being. By guiding the evaluation of existing systems and the integration of cognitive ergonomic considerations, the DEJI model guides organizations to create adaptive, human-centered solutions that address the demands of modern manufacturing environments.

In the context of additive manufacturing, the DEJI model can facilitate the seamless integration of cognitive ergonomics by addressing key challenges such as:

- interface complexity.
- decision-making under time constraints.
- task variability.

Through its structured evaluation and implementation process, the model ensures that cognitive demands are balanced with system capabilities, ultimately driving efficiency and innovation in AM practices. This chapter explores the use of the DEJI model and how it provides a roadmap for effectively designing work systems and incorporating cognitive ergonomics into work systems like additive manufacturing.

STAGES OF THE DEJI MODEL

The DEJI model of systems engineering provides one additional option for systems engineering development applications. Although the model is generally applicable in all types of systems modeling, systems quality is specifically used to describe how the DEJI model is applied. The core stages of the DEJI model are:

- Design.
- Evaluation.
- Justification.
- Integration.

Design encompasses any system initiative providing a starting point for a project. Thus, design can include technical product design, process initiation, and concept development. In essence, we can say that "design" represents requirements and specifications. Evaluation can use a variety of metrics both qualitative and quantitative, depending on the organization's needs. Justification can be based on monetary, technical, or social reasons. Integration needs to be done with respect to the normal or standard operations of the organization. Figure 10.1 illustrates the full profile of the DEJI model.

This chapter advocates using the DEJI Systems Model for technology and systems integration. An application guide is presented in a structural template for systems design, evaluation, justification, and integration. System quality is used as the basis for illustrating how the model is applied across its four stages. Specific emphasis is

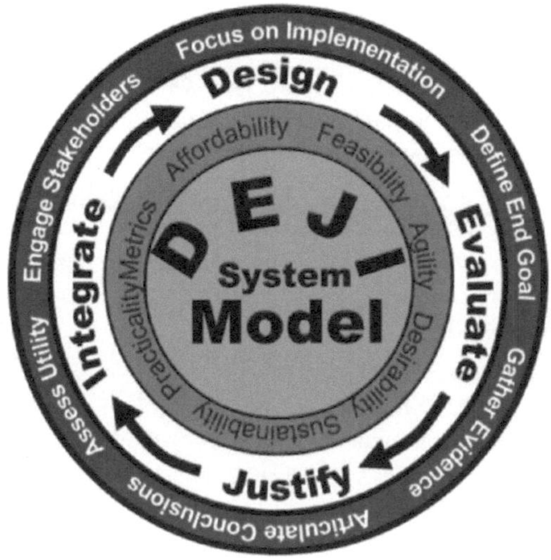

FIGURE 10.1 The DEJI systems model.

placed on applying the model to technology management in the context of pursuing and tackling UN's, SDGs, and NAE's Grand Challenges.

All the operational elements embedded in the DEJI model are explained and described as presented below:

- Design embodies Agility, Define End Goal, and Engage Stakeholder.
- Evaluate embodies Feasibility, Metrics, Gather Evidence, and Assess Utility.
- Justify embodies Desirability, Focus on Implementation, and Articulate Conclusions.
- Integrate embodies Affordability, Sustainability, and Practicality.

For application purposes, these elements interface and interact systematically to enhance the overall operational performance of the organization.

APPLICATION TO WORKFORCE QUALITY SYSTEM

Several aspects of quality must undergo rigorous research along the realms of both quantitative and qualitative characteristics. Many times, quality is taken for granted and the flaws only come out during the implementation stage, which may be too late to rectify. The growing trend in product recalls is a symptom of a priori analysis of the sources and implications of quality at the product conception stage. This column advocates the use of the DEJI Model for enhancing quality design, quality evaluation, quality justification, and quality integration through hierarchical and stage-by-stage processes.

Incorporating cognitive ergonomics into work system design can further enhance quality by addressing operator cognitive load, decision-making processes, and task variability. Organizations can create more efficient and user-friendly systems that support consistent quality outcomes by aligning system capabilities with human cognitive strengths and limitations.

Better quality is achievable, and there is always room for improvement in the quality of products and services. However, we must commit more efforts to research at the outset of the product development cycle. Even the human elements of the perception of quality can benefit from more directed research from a social and behavioral sciences point of view. The DEJI model prompts system designers to consider human factors to improve the quality of the work system design.

> **Performance Accountability**
>
> Throughout history, engineering has answered the call of society to address specific challenges. (Kirby et al, 1990; Grayson, 1993) With such answers comes a greater expectation of professional accountability. Consider the level of social responsibility that existed during the time of the Code of Hammurabi. (Bryant, 2005) Two of the laws are echoed below:
>
> **Hammurabi's Law 229**:
>
> "If a builder builds a house for someone, and does not construct it properly, and the house which he built falls in and kills its owner, then that builder shall be put to death."
>
> **Hammurabi's Law 230:**
>
> "If it kills the son of the owner, the son of that builder shall be put to death."

These are drastic measures designed to curb professional dereliction of duty and enforce social responsibility with particular focus on product quality. Research and education must play bigger and more direct roles in the design, practice, and management of quality. Duckworth and Moore (2010) present modern aspects of social responsibility in the context of day-to-day personal and professional activities. The global responsibility of the greater society is presented by Bhargava (2006) with respect to world development challenges covering the global economy, human development, global governance, and social relationships. Quality is a common theme in development challenges. Focusing on the emerging field of Big Data, Xian and Madhavan (2014) advocate for engineering education collaboration, which aligns well with data-intensive product development. With the above principles as possible tenets for better research, education, and practice of quality in engineering and technology, this column presents the DEJI model as a potential methodology.

THE DEJI SYSTEMS MODEL

The DEJI (Design, Evaluation, Justification, and Integration) model encourages the practice of building quality into a product right from the beginning so that the product integration stage can be more successful. Figure 10.2 shows the four-legged model for an overall systems quality with respect to design, evaluation, justification, and integration. Table 10.1 shows the potential elements within each stage of the DEJI model.

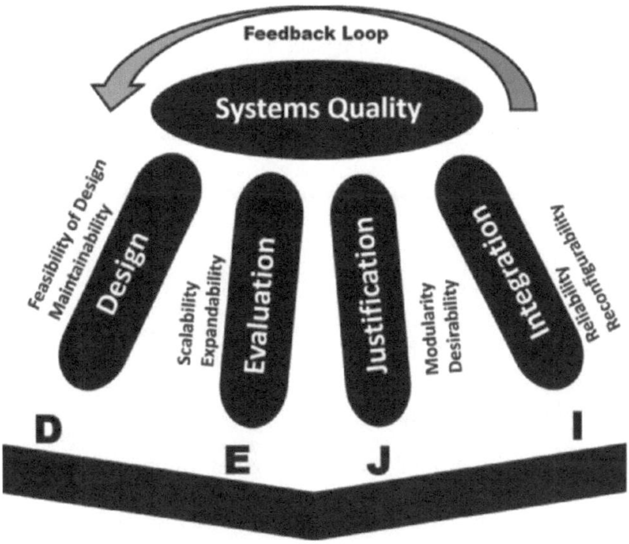

FIGURE 10.2 Footstool elements of the DEJI systems model.

TABLE 10.1
Tabulated Layout of DEJI Model

DEJI Model	Characteristics	Tools and Techniques
Design	Define goals.	Parametric assessment.
	Set performance metrics.	Project state transition.
	Identify milestones.	Value stream analysis.
Evaluate	Measure parameters.	Pareto distribution.
	Assess attributes.	Life cycle analysis.
	Benchmark results.	Risk assessment.
Justify	Assess economics.	Benefit-cost ratio.
	Assess technical output.	Payback period.
	Align with goals.	Present value.
Integrate	Embed in normal operation.	SMART concept.
	Verify symbiosis.	Process improvement.
	Leverage synergy.	Quality control.

THE DESIGN PHASE

The design of quality in product development should be structured to follow point-to-point transformations. A good technique to accomplish this is the use of state-space transformation, with which we can track the evolution of a product from the concept stage to final product stage.

Design, in the context of the DEJI Systems Model, is not limited to just the physical design of a product. Design encompasses a collaborative template across multiple factors, including the following:

- Legal considerations.
- Regulatory requirements.
- Safety concerns.
- Reliability expectations.
- Sustainability mandate.
- Environmental impact.
- Lifecycle cost reduction.
- Optimization across platforms.
- Maintainability.
- Human-system integration.

For technology quality design, the following definitions are applicable:

Product state: A state is a set of conditions that describe the product at a specified point in time. The *state* of a product refers to a performance characteristic of the product which relates input to output such that a knowledge of the input function over time and the state of the product at time $t = t_0$ determines the expected output for $t \geq t_0$. This is particularly important for assessing where the product stands in the context of new technological developments and the prevailing operating environment.

Product state space: A product *state space* is the set of all possible states of the product lifecycle. State space representation can solve product design problems by moving from an initial state to another state, and eventually to the desired end-goal state. The movement from state to state is achieved by means of actions. A goal is a description of an intended state that has not yet been achieved. The process of solving a product problem involves finding a sequence of actions that represents a solution path from the initial state to the goal state. A state space model consists of state variables that describe the prevailing condition of the product. The state variables are related to inputs by mathematical relationships. Examples of potential product state variables include schedule, output quality, cost, due date, resource, resource utilization, operational efficiency, productivity throughput, and technology alignment. For a product described by a system of components, the state-space representation can follow the quantitative metric below:

$$Z = f(z, x); \; Y = g(z, x)$$

Where f and g are vector-valued functions. The variable Y is the output vector while the variable x denotes the inputs. The state vector Z is an intermediate vector relating x to y. In generic terms, a product is transformed from one state to another by a driving function that produces a transitional relationship given by:

$$S_s = f(x \mid S_p) + e,$$

Where S_s = subsequent state; x = state variable; S_p = the preceding state; e = error component.

The function *f* is composed of a given action (or a set of actions) applied to the product. Each intermediate state may represent a significant milestone in the project. Thus, a descriptive state space model facilitates an analysis of what actions to apply to achieve the next desired product state. Figure 10.3 shows a representation of a product development example involving the transformation of a product from one state to another through the application of human or machine actions. This simple representation can be expanded to cover several components within the product information framework. Hierarchical linking of product elements provides an expanded transformation structure. The product state can be expanded in accordance with implicit requirements. These requirements might include grouping of design elements, linking precedence requirements (both technical and procedural), adapting to new technology developments, following required communication links, and accomplishing reporting requirements. The actions to be taken at each state depend on the prevailing product conditions. The nature of subsequent alternate states depends on what actions are implemented. Sometimes there are multiple paths that can lead to the desired end result. At other times, there exists only one unique path to the desired objective. In conventional practice, the characteristics of the future states can only be recognized after the fact, thus making it impossible to develop adaptive plans. In the implementation of the **DEJI** model, adaptive plans can be achieved because the events occurring within and outside the product state boundaries can be considered.

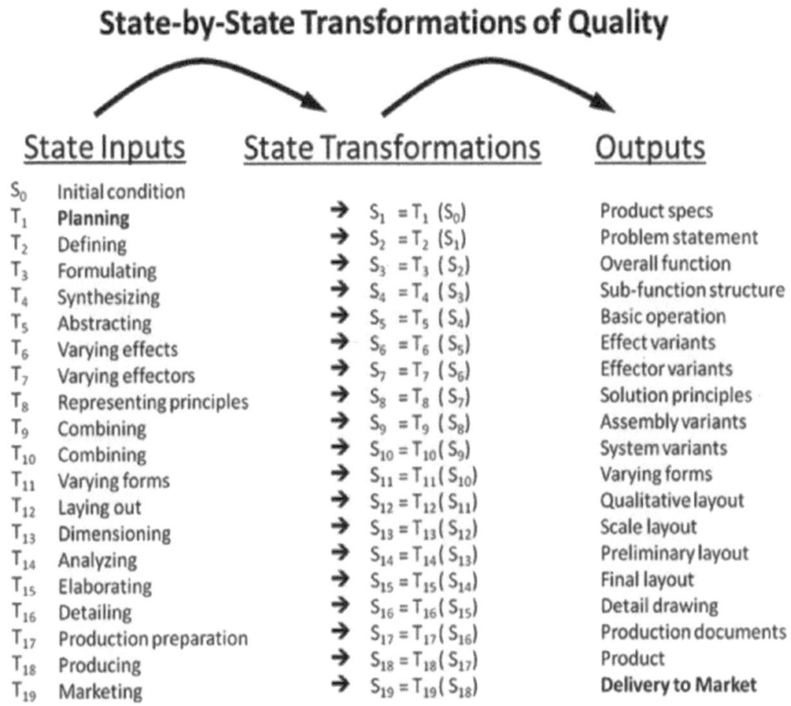

FIGURE 10.3 State-Space transformations of quality.

If we describe a product by P state variables s_i, then the composite state of the product at any given time can be represented by a vector **S** containing P elements. That is,

$$\mathbf{S} = \{s_1, s_2, ..., s_P\}$$

The components of the state vector could represent either quantitative or qualitative variables (e.g., cost, energy, color, time). We can visualize every state vector as a point in the state space of the product. The representation is unique since every state vector corresponds to one and only one point in the state space. Suppose we have a set of actions (transformation agents) that we can apply to the product information to change it from one state to another within the project state space. The transformation will change a state vector into another state vector. A transformation may be a change in raw material or a change in design approach. The number of transformations available for a product characteristic may be finite or unlimited. We can construct trajectories that describe the potential states of a product's evolution as we apply successive transformations with respect to technology forecasts. Each transformation may be repeated as many times as needed. Given an initial state $\mathbf{S}_0$, the sequence of state vectors is represented by the following:

$$\mathbf{S}_n = T_n(\mathbf{S}_{n-1}).$$

The state-by-state transformations are then represented as $\mathbf{S}_1 = T_1(\mathbf{S}_0)$; $\mathbf{S}_2 = T_2(\mathbf{S}_1)$; $\mathbf{S}_3 = T_3(\mathbf{S}_2)$; .. . . . ; $\mathbf{S}_n = T_n(\mathbf{S}_{n-1})$. The final State, $\mathbf{S}_n$, depends on the initial state **S** and the effects of the actions applied.

EVALUATION OF WORKFORCE QUALITY

A product can be evaluated based on cost, quality, schedule, and meeting requirements. There are many quantitative metrics that can be used in evaluating a product at this stage. Learning curve productivity is one relevant technique that can be used because it offers an evaluation basis of a product with respect to the concept of growth and decay. The half-life extension (Badiru, 2012) of basic learning is directly applicable because the half-life of the technologies going into a product can be considered. In today's technology-based operations, retention of learning may be threatened by fast-paced shifts in operating requirements. Thus, it is of interest to evaluate the half-life properties of new technologies as the impact on the overall product quality. Information about the half-life can tell us something about the sustainability of learning-induced technology performance. This is particularly useful for designing products whose life cycles stretch into the future in a high-tech environment.

JUSTIFICATION OF QUALITY

We need to justify a program based on quantitative value assessment. The Systems Value Model (SVM) is a good quantitative technique that can be used here for project justification based on value. The model provides a heuristic decision aid for comparing project alternatives. It is presented here again for the present context.

Value is represented as a deterministic vector function that indicates the value of tangible and intangible attributes that characterize the project. It is represented as $V = f(A, A_2, \ldots, A_p)$, where V is the assessed value, and the A values are quantitative measures or attributes. Examples of product attributes are quality, throughput, manufacturability, capability, modularity, reliability, interchangeability, efficiency, and cost performance. Attributes are a combined function of factors. Examples of product factors are market share, flexibility, user acceptance, capacity utilization, safety, and design functionality. Factors are themselves considered to be composed of indicators. Examples of indicators are debt ratio, acquisition volume, product responsiveness, substitutability, lead time, learning curve, and scrap volume. By combining the above definitions, a composite measure of the operational value of a product can be quantitatively assessed. In addition to the quantifiable factors, attributes, and indicators that impinge upon overall project value, the human-based subtle factors should also be included in assessing overall project value.

EARNED VALUE TECHNIQUE

Value is synonymous with quality. Thus, the contemporary earned value technique is relevant for "earned quality" analysis. This is a good analytical technique to use for the justification stage of the DEJI model. This will impact cost, quality, and schedule elements of product development with respect to value creation. The technique involves developing important diagnostic values for each schedule activity, work package, or control element as shown in Figure 10.4. The variables in the figure

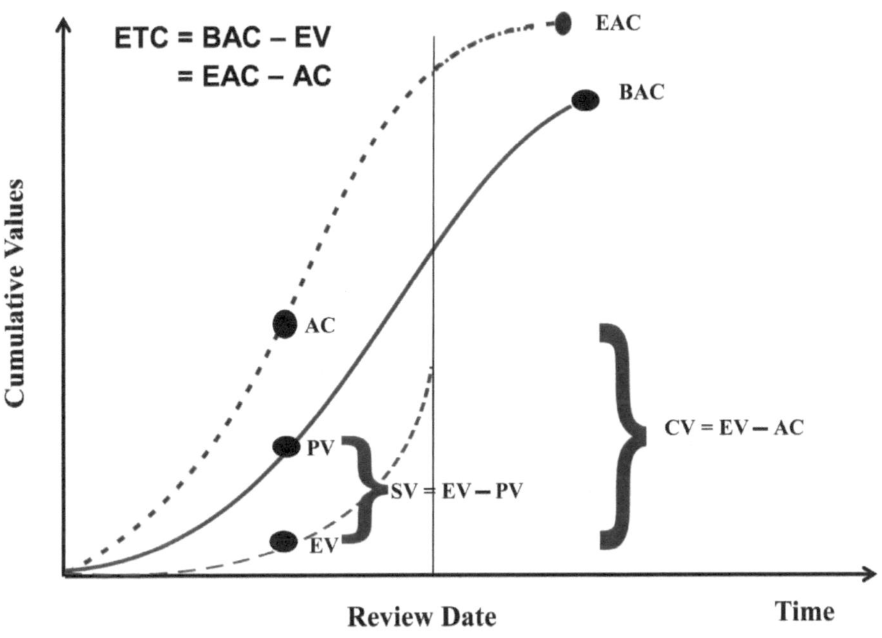

FIGURE 10.4 Earned value technique for quality assessment.

are PV: Planned Value; EV: Earned Value; AC: Actual Cost; CV: Cost Variance; SV: Schedule Variance; EAC: Estimate at Completion; BAC: Budget at Completion; ETC: Estimate to Complete. This analogical relationship is a variable research topic for quality engineering and technology applications.

INTEGRATION OF WORKFORCE PERFORMANCE

Without being integrated, a system will be in isolation and it may be worthless. We must integrate all the elements of a system based on the alignment of functional goals. The overlap of systems for integration purposes can conceptually be viewed as projection integrals by considering areas bounded by the common elements of subsystems as shown in Figure 10.5. Quantitative metrics can be applied at this stage for effective assessment of the product state. Trade-off analysis is essential in quality integration. Pertinent questions include the following:

- What level of trade-offs on the level of quality are tolerable?
- What is the incremental cost of higher quality?
- What is the marginal value of higher quality?
- What is the adverse impact of a decrease in quality?
- What is the integration of quality of time? In this respect, an integral of the form below may be suitable for further research:

$$I = \int_{t_1}^{t_2} f(q)dq$$

Where I = integrated value of quality, $f(q)$ = functional definition of quality, t_1 = initial time, and t_2 = final time within the planning horizon.

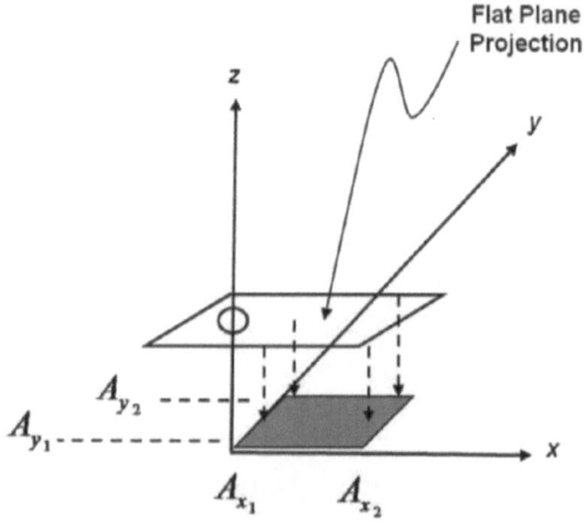

FIGURE 10.5 Technology integration surfaces

Presented below are guidelines and important questions relevant to quality integration.

- What are the unique characteristics of each component in the integrated system?
- How do the characteristics complement one another?
- What physical interfaces exist among the components?
- What data/information interfaces exist among the components?
- What ideological differences exist among the components?
- What are the data flow requirements for the components?
- What internal and external factors are expected to influence the integrated system?
- What are the relative priorities assigned to each component of the integrated system?
- What are the strengths and weaknesses of the integrated system?
- What resources are necessary to keep the integrated system operating satisfactorily?
- Which organizational unit has primary responsibility for the integrated system?

The systems approach of the DEJI model will facilitate a better alignment of product technology with future development and needs. The stages of the model require research for each new product with respect to design, evaluation, justification, and integration. Existing analytical tools and techniques, as well as other systems engineering models, can be used at each stage of the model. Thus, a hybrid systems modeling is possible. Quality is an integrative process that must be evaluated on a stage-by-stage approach. This requires research, education, and implementation strategies that consider several pertinent factors. This column suggests the DEJI model, which has been used successfully for product development applications, as a viable methodology for quality design, quality evaluation, quality justification, and quality integration. This column is intended as a source to spark the interest of researchers to apply this tool in new product development efforts.

APPLICATION GUIDE FOR SDGs AND GRAND CHALLENGES

A template for applying the DEJI Systems Model to workforce development challenges can be summed up in Figure 10.6. The dimensions of applications of the DEJI Systems Model cover the following elements:

- Productivity.
- Efficiency.
- Effectiveness.
- Integration.

The interweaving ovals in the figure signify a non-discrete relationship among the Design, Evaluation, Justification, and Integration elements of the model. The user will determine what factors or indicators fall within each stage of implementing the model and how those factors interrelate to other factors in the other stages.

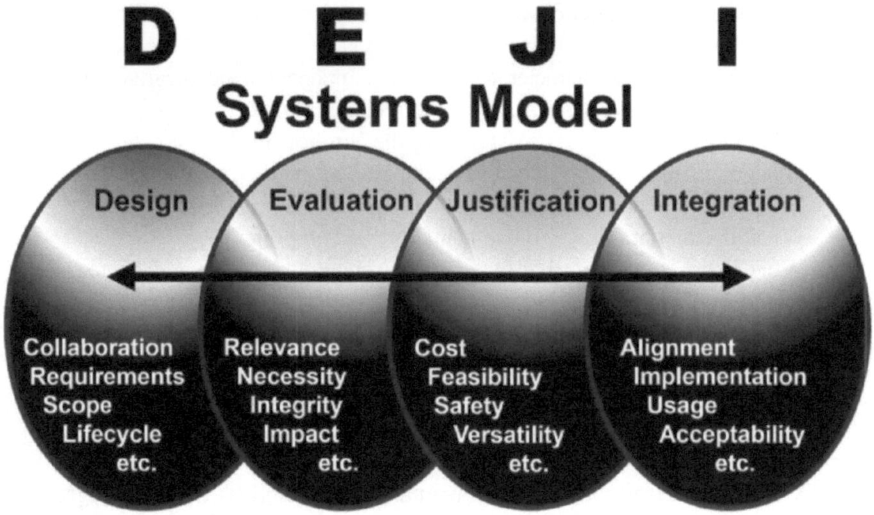

FIGURE 10.6 DEJI systems model application ovals.

In the context of the model, design implies the specification of requirements. Without well-designed, well-defined, and well-articulated requirements, all subsequent efforts will be misguided. The modular components represent plug-and-play properties, implying systems interchangeability. DEJI Systems Model facilitates system modularity. Items designed in modules tend to offer more flexibility and adaptability. This is why industrial engineering is a versatile approach to system implementations of UN's, SDG's, and NAE's Grand Challenges. With the structured framework of the DEJI Systems Model, technology implementations can be executed faster, more efficiently, effectively, and economically. Integration is the hallmark of the model. Integration should be done within the local context (alignment with prevailing scenarios) to maximize the possibility of success. Organizations are continuing to insert this highly functional framework into nearly every aspect of modern technology management. Everything is connected and very interdependent now, so everything now requires a systems engineering approach. DEJI is a flagship framework for planning contemporary projects, systems, teams, and technology ventures. Many organizations are excited about a lot of things related to new technology, but they don't have the foggiest idea how to implement and integrate technology into their existing operations seamlessly. For example, safety by design is enabled by the model's structural methodology.

Workforce Work Design Quote

Design is not how it looks and feels, it's how it works.

Steve Jobs

"How it works" is a function of how it integrates with other work elements within the organization. In this sense, the design or specifications of the UN, SDGs, and NAE's Grand Challenges must align with the prevailing situations in each target area.

UTILIZATION OF THE DEJI MODEL TO IMPROVE WORK SYSTEM DESIGN QUALITY IN ADDITIVE MANUFACTURING

Additive manufacturing (AM) is a fabrication process that creates three-dimensional products from computer-aided design by depositing or curing material layer by layer. Usually, it uses lasers to melt or solidify material, ensuring precision and accuracy in building complex geometries. Traditionally, AM has been used for prototyping, producing complex parts, spare parts, and small lot production (Roscoe et al, 2023). As additive manufacturing (AM) technology advances, its integration into mass production is set to redefine the manufacturing landscape, creating new opportunities and challenges for industries worldwide. Recent developments are driving a shift in AM's role, moving beyond prototyping to full-scale rapid manufacturing. These advancements include material improvements, technological innovations, and cost reductions. Material advancements have enhanced AM-produced parts' strength and durability, making them comparable to those created through traditional manufacturing methods. Technological innovations have further accelerated AM adoption by making processes faster, more precise, and more reliable. At the same time, the declining costs of AM machines and materials have significantly reduced barriers to entry, enabling more businesses to adopt the technology. Together, these developments are expanding AM's capabilities while requiring a reimagining of workforce development and work design strategies to capitalize on its potential fully. The human-in-the-loop must be considered to fully capitalize on its potential, including human factors such as cognitive ergonomics, as discussed in Chapter 5. AM requires more cognitive ergonomics analysis than traditional manufacturing processes due to its constant use of computer interfaces.

The strategy discussed below for utilizing the DEJI model in human factors for additive manufacturing is adapted from Doran et al (2022). We will discuss applying the equations previously explained in this chapter to improve AM's work system design quality. The process begins with the DESIGN phase, which involves defining the product's state at specific points in time. The focus is on assessing the operator's state during various stages of the 3D printing process. These stages include:

- Conceptualization of the 3D model.
- Review, revision, and finalization of the 3D model.
- Setup for printing the 3D model.
- Monitoring the product during the printing stage.
- Cleaning the 3D printer and organizing excess powder post-printing.
- Filtering the excess powder.
- Post-process analysis of the product.
- Assess cognitive work elements.

At each of these specified stages, an analysis can be conducted using the quantitative metric:

$$Z = f(z, x); \; Y = g(z, x)$$

To evaluate the input and expected output. This analysis considers state variables such as product cost, final product due date, output quality, throughput, resource utilization, and operational efficiency.

In the additive manufacturing process, $S_s = f(x \mid S_p) + e$ is then used to measure the transitional relationships between product states. The state inputs for this process include:

- Planning.
- Defining.
- Designing.
- Revising.
- Finalizing.
- Printing.
- Filtering.
- Cleaning.
- Machining.
- Analyzing.

This analysis identifies the actions required to transition to the next product state.

For a deeper product analysis, $\mathbf{S} = \{s_1, s_2, \ldots, s_P\}$ can describe the potential states of a 3D-printed part, transitioning from inputs like a 3D-modeled part or raw material (e.g., metal powder or plastic) to the final printed product.

Lastly, $\mathbf{S}_n = T_n(\mathbf{S}_{n-1})$ monitors state-by-state transformations by analyzing the state inputs and outputs. The outputs include Product Specifications, Problem Statement, 3D Part Layout, Revised 3D Part Layout, Final 3D Part Layout, Machine Setup, Fabrication, Reusable Material, Prepared Machine, Finalized Part, and Product Complete. With these equations calculated, the Evaluation phase follows.

For human factors in additive manufacturing, the EVALUATION phase assesses the operator's efficiency in printing a quality part and the time required to complete tasks throughout the 3D printing process. These evaluations focus on two primary metrics:

- Operator performance and efficiency.
- Mental workload during the 3D printing tasks.

By analyzing these metrics, human factors issue within additive manufacturing can be identified, revealing tasks that cause inefficiencies or heightened cognitive demands. Solutions such as improved task design or the integration of AI tools can then be explored to mitigate these issues. The next step involves ensuring that these tasks are value-added activities, leading into the Justification phase.

The JUSTIFICATION phase involves validating the necessity and value of each task in the 3D printing process, ensuring they contribute tangible or intangible benefits. $V = f(A, A_2, \ldots, A_p)$ This is applied to assess the value of tasks using quantitative measures like operator efficiency, reliability, and part quality. Additional justifications pertain to safety precautions, such as the use of Personal Protective Equipment (PPE) to minimize static and oxygen monitoring during the room filtration process. These justifications are crucial to validating the DEJI model's implementation for human factors in additive manufacturing.

The final stage, INTEGRATION, ensures a seamless application of the DEJI model within the existing safety protocols of additive manufacturing. The 3D printing process revolves around the operator's ability to design effective 3D models and prepare the printer to produce functional replicas. Therefore, maximizing human performance is paramount. Quantitative metrics such as cost and resource utilization, previously used in product state assessments, can further inform this phase. By following this strategy, the DEJI model's application for human factors in additive manufacturing ensures the process is safe, user-friendly, and effective, enhancing productivity outcomes.

SUMMARY

This chapter explores the DEJI model and its application to additive manufacturing to enhance work design quality. It emphasizes the importance of incorporating cognitive considerations in work design to support human operators better and increase process efficiency. Advancing cognitive ergonomics in additive manufacturing can lead to more efficient, safer, and human-centered work environments, ultimately enhancing individual and organizational performance. Further research is necessary to advance the integration of cognitive ergonomics in modern manufacturing systems, and the DEJI model is a methodology to guide future work for improved system design. See Badiru (2013), Badiru (2019), Gao et al (2015), Miller (2010) and Wohlers and Caffrey (2023) for related topics.

REFERENCES

Badiru, A. B. (2012), "Application of the DEJI Model for Aerospace Product Integration," ***Journal of Aviation and Aerospace Perspectives (JAAP)***, Vol. 2, No. 2, pp. 20–34, Fall 2012.

Badiru, A. B. (November 2013), "Up to the Task: Industrial Engineering Is Essential to Meeting NAE's Grand Challenges," ***Industrial Engineer***, Vol. 45, No. 11, pp. 42–45.

Badiru, A. B. (2019), **Systems Engineering Models: Theory, Methods, and Applications**, Taylor & Francis/CRC Press, Boca Raton, FL.

Bhargava, V. editor (2006), **Global Issues for Global Citizens: An Introduction to Key Development Challenges**, The World Bank, Washington, DC.

Bryant, T. (2005), **The Life & Times of Hammurabi**, Mitchell Lane Publishers, Newark, DE.

Doran, E., S. Bommer, and A. Badiru (2022), "Integration of Human Factors, Cognitive Ergonomics, and Artificial Intelligence in the Human-Machine Interface for Additive Manufacturing," ***International Journal of Mechatronics and Manufacturing Systems***, Vol. 15, No. 4, pp. 310–330.

Duckworth, H. A., and R. A. Moore (2010), **Social Responsibility: Failure Mode Effects and Analysis**, Taylor and Francis/CRC Press, Boca Raton, FL.

Gao, W., Y. Zhang, D. Ramanujan, et al. (2015), "The Status, Challenges, and Future of Additive Manufacturing in Engineering," ***Computer-Aided Design***, Vol. 69, pp. 65–89.

Grayson, L. P. (1993), **The Making of an Engineer: An Illustrated History of Engineering Education in the United States and Canada**, John Wiley, New York, NY.

Kirby, R. S., S. Withington, A. B. Darlin, and F. G. Kilgour (1990), **Engineering in History**, Dover Publications, New York, NY.

Miller, R. K. (2010), "The Educational Imperatives of the Engineering Grand Challenges," Keynote Address, ASEE Global Colloquium on Engineering Education, Singapore, October 19–21.

Roscoe, S., P. D. Cousins, and R. Handfield (2023), "Transitioning Additive Manufacturing from Rapid Prototyping to High-Volume Production: A Case Study of Complex Final Products," ***Journal of Product Innovation Management***, Vol. 40, No. 4, pp. 554–576.

Wohlers, T., and T. Caffrey (2023), ***Wohlers Report 2023: 3D Printing and Additive Manufacturing State of the Industry***, Wohlers Associates, Fort Collins.

Xian, H., and K. Madhavan (2014), "Anatomy of Scholarly Collaboration in Engineering Education: A Big-Data Bibliometric Analysis," ***Journal of Engineering Education***, Vol. 103, No. 3, pp. 486–514.

11 The Future of Work and Workforce Development in an Era of Artificial Intelligence

INTRODUCTION

There is no escaping the emergence of Artificial Intelligence (AI) in the workforce. Technology is rapidly evolving every day, and whether we like it or not, AI will eventually become an integral part of the workforce, even more so than it already is. AI is an extremely useful tool that can aid in all areas of the workplace including business, employee, and customer/client relations. While it's commonly believed that AI is a relatively new technology, this is actually far from the truth. AI, or at least automated technology, has existed for decades. However, the development of AI has skyrocketed within the last century, notably beginning in the 1950s. Over the course of several decades, AI has developed rapidly and only continues to evolve and become more and more useful. There are many different types of AI that can be integrated into the workplace including Generative AI, Limited Memory AI, and Machine Learning. These different types of AI can enhance the workplace for employers, employees, and customers because they are able to handle tedious tasks, provide quick and accessible information, and imitate human connections. AI can also aid in skill retention, allowing employees to better understand their work, as well as provide them with opportunities to learn new skills and information. While there are several advantages to integrating AI into the workplace, there are also several disadvantages. It is important for companies to be strategic about incorporating AI into these business practices because it not only affects them but also their employees and customers as well.

AI will no doubt aid in worker and workforce development. These technologies will certainly change many industries for the better, so long as companies use them efficiently and ethically.

HISTORY OF ARTIFICIAL INTELLIGENCE

Artificial Intelligence is a field of computer science that uses technology to create machines and systems capable of performing tasks that typically require human intelligence. The idea of AI dates back to ancient times, when inventors created machines that moved independently without human intervention called "automatons." One of the most famous automatons was created by renowned Italian polymath Leonardo

DOI: 10.1201/9781003537755-11

da Vinci. Moving forward, interest in AI came to a head in the 1950s, when Alan Turing published "Computer Machinery and Intelligence." This eventually became known as the Turing Test, which was used to measure computer intelligence. The term "artificial intelligence" was coined by John McCarthy, who held a workshop at Dartmouth University about the topic. Between the mid-1950s and the late 1970s, AI research and development grew rapidly. There were times when the research and development of AI went through periods called "AI winters," where funding and interest in the topic decreased, making further research significantly more difficult. There are several milestones throughout the last century depicting developments in the research and advancement of AI.

In 1958, John McCarthy created List Processing (LISP), the first programming language for AI research, which is still used to this day. In 1961, the first industrial robot, called Unimate, was created and began working on an assembly line for General Motors in New Jersey. The first chatbot (known then as "chatterbot") was invented in 1966 by Joseph Weizenbaum. It was called ELIZA and served as a mock psychotherapist that used natural language processing to converse with human participants. By 1979, the Association for the Advancement of Artificial Intelligence (AAAI) was established.

There was a significant growth of interest in the development of AI in the 1980s known as the "AI boom." There were major breakthroughs in research and development, as well as an increase in government funding to support researchers. The first AAAI conference was held at Stanford University and XCON, the first expert system, reached the commercial market. America was not the only country to join the AI boom. The Japanese government allocated $850 million in 1981 for their project, the Fifth Generation Computer, which aimed to create computers that could translate, converse, and express reasoning on a human level. In Germany, Ernst Dickmann and his team of researchers at Bundeswehr University of Munich created the first driverless car, which could drive up to 55 mph on roads without obstacles or other human drivers. By 1987, Alacrity, the first strategy managerial advisory system by Alacritous, Inc., was launched commercially.

Things came to a halt in 1987 when an AI winter came. The term AI winter describes a period of low consumer, public, and private interest in the research and development of AI, leading to a decrease in funding and few breakthroughs. This AI winter was caused by setbacks in the machine market and expert systems, including the end of the Fifth Generation project. Despite the lack of funding and breakthroughs during the AI winter, the 1990s and 2000s showed promise in the research and development of AI. Between 1993 and 2011, several technological advancements were made, many of which include popular products and technologies that are used in the present day. In 1997, Deep Blue was developed by IBM and became the first computer program to beat a human world chess champion. That same year, Windows released a speech recognition software. In 2002, the first Roomba, an autonomous vacuum that navigates around clutter and beneath furniture to clean floors and returns to its charger upon completion, was released for consumption. The next year, in 2003, NASA landed two rovers, Spirit and Opportunity, on Mars, which were able to navigate the surface of the planet without human intervention. In 2006, media platforms such as Twitter, Facebook, and Netflix began utilizing AI as part of their advertising campaigns and user experience algorithms. In 2010, Microsoft launched the Xbox 360 Kinect, the first gaming hardware designed to track body movement and translate it into gaming directions. There were major breakthroughs made the next year in 2011. IBM launched Watson, an NLP computer programmed to answer questions, and Apple released Siri, the first popular virtual assistant, easily accessible on all of their devices.

Since the pivotal launch of Siri, research and development of AI skyrocketed in 2012 and has evolved significantly, continuing to evolve further in the present day. In 2012, two researchers from Google trained a neural network to recognize cats based on unlabeled images without any background information. In 2015, thousands of major tech moguls, including Elon Musk and Stephen Hawking, signed an open letter to the world's government systems which banned the development and use of autonomous weapons for war purposes. The first humanoid robot, named Sophia, was created in 2016 by Hanson Robotics. Sophia was the first robot created with realistic human features and attributes. It also had the capability to see, communicate, and replicate human emotions. In 2017, Facebook programmed two AI chatbots to converse and negotiate in English, but the programs ended up forgoing the use of English and created their own language autonomously. Later the next year, Alibaba, a Chinese tech group, created a language-processing AI system that was able to beat human intellect on a Stanford University reading and comprehension test. In 2020, OpenAI started beta testing ChatGPT-3, a generative AI model that uses Deep Learning to create code, poetry, and other language and writing tasks. There have been significant improvements made to ChatGPT, with its most current version being known as ChatGPT 4o mini. This generative AI model, along with other models, including Google's Gemini, can work with prompts entered by human users. These systems can quickly respond to whatever a human user prompts them with and provide better information when well-crafted prompts are input. DALL-E, another generative AI system developed by OpenAI in 2021, can process and understand images enough to produce accurate captions, further accelerating AI's ability to understanding the visual world.

The future holds incredible potential for the advancement of current AI models and systems, as well as the development of new ones. We can also expect to see the research and development of AI in the workforce and niche fields including the medical, legal, and engineering fields.

TYPES OF AI SYSTEMS USED IN THE WORKPLACE

There are several types of AI that can be used to enhance different areas in the workforce. From finding useful information quickly to providing enhanced learning experiences for employee training, AI can be beneficial to all industries. Some types of AI that can be beneficial to the workforce include Reactive Machine AI, Limited Memory AI, Machine Learning, and Generative AI (GenAI). Definitions for these different types of AI and what they can be used for are listed below

- **Reactive Machine AI:** Systems with no memory that are designed to perform very specific tasks. Since these systems cannot recollect previous outcomes or decisions, they only work with presently available data. This type of AI model stems from statistical math and can analyze vast amounts of data in order to produce a seemingly intelligent output. (IBM) Examples of this type of AI in the workforce include IBM Deep Blue and Netflix's Recommendation Engine. IBM Deep Blue is IBM's chess-playing supercomputer which was able to beat chess world champion Garry Kasparov in the late 1990s. Netflix's Recommendation Engine offers recommendations to viewers based on what they previously watched. It analyzes this data and puts together a list of content viewers are more likely to enjoy based on genre and thematic elements of previously watched content, which can make it easier for viewers to find something they'll enjoy.

- **Limited Memory AI:** These AI systems are able to recall past events and outcomes, as well as monitor specific objects and situations over time, unlike its predecessor Reactive Machine AI. It can use past and present data to decide on a course of action that will work best for a desired outcome. While this type of AI can retain information for a specific amount of time, it cannot store that data in a library of information to be used over a long period of time. Some examples of this type of AI include GenAI tools such as ChatGPT (OpenAI), Bard, Gemini, and DeepAI. These systems rely on the capabilities of Limited Memory AI to predict what comes next within the content they are generating. Other examples of Limited Memory AI include virtual assistants and chatbots such as Apple's Siri, Amazon's Alexa, and IBM Watson. Self-driving cars also use Limited Memory AI to be able to understand the world around them and make informed decisions about standard driving practices.
- **Machine Learning (ML):** Machine Learning is a specific subset of AI and refers to algorithms that teach machines to learn from data. (Office Space) These machines are able to learn without being given any explicit programming. Machine Learning can be very useful in the workforce, especially for companies and businesses searching for ways to better automation.
- **Generative AI (GenAI):** A new subset of AI that's able to create content, such as generating an image, design, or text. GenAI models learn the underlying patterns and structures of their training data fand use them to produce new data based on the input, often in the form of natural language prompts from human users. GenAI models such as ChatGPT, Bard, and Gemini can write poems, stories, essays, letters, etc., based on any given prompt, and it can also provide suggestions for queries by using information it finds about the subject of the prompt. Human Resource (HR) representatives can use GenAI to create personalized training programs for new employees. (AMA)

All of these systems can be useful in different areas of the workplace. Many industries including tech, medical, and retail have employed the use of Virtual Assistants when customers call their customer service lines. Through prompts, the virtual assistant can direct people to human representatives. Many people feel this is an arduous task, as many times the robot does not understand the conversation. While this sometimes may be the case, there's no doubt that AI proves to be integral in the workforce.

AI AND THE FUTURE OF WORK

The incorporation of AI in the workforce provides learning opportunities for new skillsets that are essential in the workforce. With the rapid evolution of AI systems, it's only a matter of time before it becomes a part of everyday life at work. According to Sarah Choudhary, CEO of Ice Innovations, there has been a rise in AI-driven jobs. She notes that her company has integrated AI into their business operations by creating new roles and transforming existing ones (Choudhary, 2024). This is something many companies will have to think about before incorporating AI into their business practices, especially because workers may be fearful of losing their jobs to AI.

It is a common fear that many believe AI will lead to mass unemployment; however, that is not the case. According to Choudhary, AI automates repetitive tasks and also augments human capabilities, which can provide workers with more time to work on strategic and creative tasks. While AI may not be capable of fully and truly understanding human emotion and empathy, it's still able to mimic human capabilities to perform tasks such as customer service queries and automated telephone lines for many different workplaces.

The tech industry is not the only industry that has shifted to the use of AI in their business practices. Some of these industries include healthcare, finance, manufacturing, and retail, all of which have adapted to using AI to improve services and products. For example, as shown in the summary in Table 11.1 (based on Isakova, 2024), the healthcare industry has incorporated AI to automate administrative tasks, provide recommendations for follow-up and lifestyle advice, and support medical research and diagnosis.

There has been a plethora of AI programs released in recent years. One of the most well-known generative AI programs is ChatGPT, created by OpenAI in 2020, captured the world's attention with its ability to produce an extensive range of seemingly human-generated content including different variations and genres of text, video, audio, images, and code (Kinder et Al, 2014). Generative AI (GenAI) is an excellent tool for automating repetitive tasks, reducing errors, and summarizing huge amounts of data in a way that's digestible and easy for human brains to process (Coveo). It can also summarize lengthy documents and resources that might take a while to read, therefore allowing more time for employees to work on other, potentially more important, projects.

It is important to consider the impact of the adoption of AI on the workforce. While all workers are impacted by this integration, it is expected that young adults, Hispanic, Native American, and Black workers will experience the highest levels of job disruption. This is due to these demographics' dramatic overrepresentation in highly automatable jobs, which are at high risk for job displacement (Roth, 2023). According to Lilian Roth, nearly thirty percent of young adult workers have high-risk jobs, where over seventy percent of the tasks could be automated. In this ever-evolving era of AI, a vast majority of these jobs could become automated. Due to these complications, it is important that employers remain supportive of all employees, especially the employees who fall under the demographics above. The integration of AI is new territory that may be met with resistance, which is why employers should meet their employees with understanding and willingness to help them adapt to these huge changes.

TABLE 11.1
Generative AI Productivity Impact

Industry Type	Investments	Dollar Impact (in Billion USD)
Pharmaceutical	• Automating drug development enhancing drug discovery • Streamlining clinical trials and enrollment • Optimizing the design and execution of clinical trials for medical devices	$110
Banking	• Offering personalized finance management advice • Transforming customer service and sales • Predicting credit risk and advancing credit scoring	$340
Education	• Providing aid in learning • Automating grading assignments • Creating personalized teaching materials and lesson plans	$230
Travel & Logistics	• Augmenting travel planning tools • Providing recommendations for travel destinations and itineraries • Optimizing traffic management systems	$300
Healthcare	• Automating administrative tasks • Providing recommendations for follow-up and lifestyle advice • Supporting medical research and diagnosis	$260
Energy	• Optimize the use of renewable energy sources • Automating and controlling energy systems	$240
Retail	• Conversational AI adoption • Providing assistance during product search and offering personalized recommendations • Generating content for marketing and sales	$390
Agriculture	• Enhancing crop management • Assisting in personalized training in agriculture • Predicting demand and supply	$70
Media & Entertainment	• Creating animation and visual effects • Supporting interactive storytelling • Producing content at-scale	$110
Insurance	• Analyzing claim data to prevent fraud • Improving risk assessments • Providing personalized product and service offerings	$70
Advanced Manufacturing	• Producing design for blueprints and instructions • Analyze data from sensors and machinery • Providing insights and decision support	$290

WORKFORCE TRAINING WITH AI

Incorporating AI into the workforce training would be an innovation and a learning experience for many workers that are unfamiliar with new technologies. It is imperative that workers keep an open mind when confronted with these new technologies, as well as remain willing to learn new systems, which may be guided by AI training technology. According to the National Fund for Workforce Solutions, the concept of "lifelong learning" is essential when it comes to AI in the workforce, especially in an era where technology is constantly changing and evolving (NFWS, 2024). While workers must be willing to adapt to this change, businesses that choose to incorporate AI into their practices must provide employees with adequate education and training on these systems. By providing resources such as AI literacy courses and visual training materials, workers will not only be able to adapt to these new technologies but also feel empowered to take on new roles in a workplace transformed by AI. Some of the ways that AI can enhance workforce training include the following:

- Basic or rudimentary explanation of concepts, ideas, and procedures
- Presentation of examples and analogies to aid understanding
- Learning motivator for concept reinforcement
- Scenario generation to mimic and practice learning objectives
- Creation of study plans for learning repetitions
- Random generation of learning quizzes and questions to aid learning
- Combination and permutation of learning objectives to simulate real world
- Automated sequencing of procedural steps
- Creation of cognitive patterns to prompt learning expectations
- Consistent activation of learning steps
- Test content generation for repeatability
- Creation of objective test assessment platforms

While there are several benefits to AI-driven training innovation, it is important to prepare for the potential downsides of this ever-evolving technology. A common fear is that AI will cause mass unemployment, and while this may be the case in some industries, its impact on this issue will not be as devastating as some people believe it will be. It's inevitable that some workers may experience job displacement or adverse effects from the incorporation of AI at their jobs; therefore, companies must be prepared to support their workers during this time of change. Educational institutions and employers should branch out and collaborate with each other to find solutions and build support systems for workers during times of job and/or career transitions. According to NWFS, applying a systems-thinking approach to this issue will allow employers to create new pathways to success for their workers while also benefiting from increased automation in the workplace.

Businesses are not the only parties responsible for ensuring the smooth transition of integrating AI into their workplaces. Policymakers also have the ability to shape how AI will affect workers – either through action or inaction (Khattar, 2023). For example, the Biden Administration in the United States was in the process of creating a National AI Strategy, which was required to contain a plan to address economic and job impacts from the use of AI. It is important for policymakers, in the U.S. and outside of it, focus on the needs of their workers as they develop and adopt AI resources in government jobs and the workforce in general.

Generative AI has been incorporated into product and catalog innovations for many businesses and corporations, and it can also be used to enhance employee training. According to Greg Hull of MD Retail and Hospitality, leaders in this industry can use cutting-edge AI tools to improve employee training through powerful content and videos, simulated experiences, and hyper-personalization (Hull, 2024). Employers can create personalized learning experiences for their employees that will be tailored to each employees' needs, allowing them to feel comfortable with the integration of these new technologies and confident in their newfound abilities.

Many workplaces have already incorporated AI into their hiring systems, HR systems, and training programs. These training programs are often generated by AI and provide an overall general education about the job and are not tailored specifically to individual employees While this may not be feasible for many companies, they should still consider ways to make their training programs more interactive for new and recurring employees to ensure they are well educated about their jobs and feel satisfied with their work life.

SKILL RETENTION AIDED BY AI

AI can be used to transform the ways people learn and retain information. Knowledge retention, the ability to store information and recall it over time, is key to effective learning. AI can enhance skill retention by offering interactive learning methods that will allow workers to retain information better than if they merely memorized information. AI can also provide personalized feedback within these interactive learning methods, which would allow workers to work on their training at their own pace or by an established deadline. Employers would be able to use this feedback to analyze their employees' learning interactions and performance levels, which would then allow them (employers) to adjust the training that is lacking in memory and skill retention. According to Welocalize, AI excels in delivering real-time, personalized feedback and adjusting to a trainee's pace and understanding, which is crucial for memory retention (Welocalize, 2024). These AI systems analyze data from employee interactions and highlight areas that need improvement, which allows them to adjust their approach and effectively retain the information they are learning.

AI offers several tools and techniques that significantly improve both the efficiency and effectiveness of corporate training programs (Welocalize, 2024). AI can transform innovative training methods including adaptive memory games and challenging recall activities, making them more adaptive, responsive, and effective. These approaches provide a more customized learning experience for employees, making training feel more engaging. By incorporating AI into these learning strategies, employees will be able to retain information easier, especially when they are given large amounts of information. Training videos are often times not interactive, which can make employees feel bored when they are required to complete these courses. Training programs like these also cater more toward short-term memory, which leads to temporary knowledge retention. By integrating active training techniques using AI, employees will build stronger connections to the learning material, allowing them to retain the information better and increasing the likelihood of being able to recall the information in the future.

According to Shay David, Co-founder, Chairman, and CEO of retrain.ai, AI can help HR leaders understand their employees' needs better, allowing them to use the knowledge acquired to enhance their (employees') careers By integrating AI into the HR side of business, talent management professionals can analyze large amounts of data, which allow them to gain insight to their employees' job satisfaction, work-life balance, skill retention, learning abilities, and career growth opportunities. This will also allow HR professionals to remain proactive should any issues arise with employees or within the company. These AI-driven insights will also give HR leaders the opportunity to present career growth opportunities to their employees based on the information provided by the AI systems. These insights can include employee achievements and skills, which will allow HR leaders to provide more growth opportunities for their employees within the company or even within another company.

Below is a condensed list of ways HR leaders can balance AI with workforce retention and productivity provided by Forbes Council Members.

1. **Understand that AI and people are not mutually exclusive**
   It is important for employers to recognize that people and AI are not mutually exclusive. Human employees and AI systems can work together to make the workforce more efficient, allowing employees to work on other tasks.
2. **Present AI as an opportunity, not a threat**
   Many employees may see the integration of AI into their workplace as a threat rather than an opportunity, which is why HR leaders should present it as the latter. In order to successfully integrate AI, employers must focus on employee relationships with the workplace and with these new technologies, and offer help if any obstacles appear.
3. **Separate repetitive tasks from ones requiring human attention and judgement**
   It is important for companies and their higher-up employees such as HR leaders and managers to separate repetitive tasks and tasks that require human attention or intervention. Tasks that can be automated should be. This would allow more time for employees to focus on important tasks and projects.
4. **Provide training and continuous learning**
   HR leaders should guide managers to use AI to augment human work rather than replace it. They should be able to identify which tasks are suitable for automation and provide employee training for new technologies. It is important for HR leaders and managers to foster a culture of continuous learning and improvement as AI technology is always evolving.
5. **Implement a phased approach to integration**
   It is important for companies and HR professionals to implement a phased approach to AI integration, so as not to overwhelm their employees.

ADVANTAGES AND DISADVANTAGES OF AI IN THE WORKPLACE

AI has become an integral part of the workplace and will continue to gain popularity as new features are released in the future. Like all new technologies, AI has both advantages and disadvantages when it comes to the workplace and its development. While AI offers a plethora of different solutions for business needs and can be a useful resource for employees, there's no denying that it has potential downfalls. Below is a list of both the advantages and disadvantages of AI in workforce development according to People Insight, an employee experience platform (Ellingrud, 2023; Byrne, 2024; Perryman, 2024; Quinstein, 2024; Lyon, 2025).

Advantages:

- **Enhanced efficiency and productivity**
  AI can be an extremely helpful tool for automating repetitive and mundane tasks that can lead employees to become bored or disengaged from their work. Using AI for these tasks would give employees more time to work on other, more complex, and/or creative tasks and projects. This not only helps the workplace boost efficiency, but it can also reduce the risk of human error in certain repetitive tasks such as data entry or scheduling.
- **Improved decision-making**
  Utilizing AI in the workplace can be used to analyze data, often times larger amounts of data at high speeds, completing the task in less time than a human employee would. AI can provide insights that can lead to better decision-making by identifying patterns and trends, which would allow managers and other higher-ups to make informed decisions based on data rather than human intuition. This could lead to better accuracy, improved levels of employee engagement, and better business outcomes.
- **Enhanced employee engagement through learning and development**
  AI can enhance employee engagement by providing personalized learning and development opportunities. AI-driven platforms and systems can identify skill gaps and also recommend tailored training programs, which would allow employees to grow and advance in their careers. This can help employees feel more confident in their work abilities, which is hugely beneficial to businesses.
- **Better work-life balance**
  AI-driven platforms and systems can aid employees in better managing their workloads, which can lead to an improved work-life balance. There are many different AI systems that can help automate routine tasks, allowing employees more time and energy to work high-priority tasks during work hours, with having to worry about it at home.
- **Enhanced customer experience**
  While AI poses many benefits for employees and businesses, it can also aid in making customer and client experiences better. AI, particularly GenAI, can improve customer interactions by providing quick and accurate responses to inquiries. Many companies in the present day have turned to using automated chatbots and virtual assistants that handle routine questions, which allows employees to focus on other things and also make sure the customer is connected to the right department or person that's able to help.

Disadvantages:

- **Potential job displacement**
  One of the most common fears about AI being incorporated in the workplace is potential job displacement. AI systems are constantly improving and becoming

more able to perform tasks that are usually carried out by humans. While AI can provide new job opportunities, there is no denying that it can, and likely will, replace human roles in some industries. It is important for companies to ensure their employees are supported and reassured during this time.

- **Lack of human touch and emotion**
  While AI has various capabilities, at the end of the day it is still a machine and therefore cannot provide the same understanding and emotion that an employee would receive from a human manager or leader. It is best practice to direct tasks that require empathy, emotional intelligence, and personal connection to human employees, who will be better suited to handle them. If businesses rely too heavily on the usage of AI, it can affect employee and company morale.
- **Ethical and privacy concerns**
  Using AI in the workplace can raise ethical and privacy concerns, especially when it comes to data collection and usage. It is important that businesses ensure that the AI systems they use are transparent, fair, and respectful of user privacy. There must be clear guidelines and policies established that address these concerns, which help maintain trust with employees, users, and/or clients.
- **Dependency on technology**
  It would be very easy for businesses to over-rely on the use of AI. This could cause issues if there were a system failure or technological issues, which are quite common, especially with newly developed technology. While technology certainly can make life easier, it is important to remember that it is not perfect and it shouldn't be heavily depended on, as many things can go wrong with it. Therefore, it is extremely importance for businesses to maintain a balance between the use of AI and human intervention, so operations can run smoothly in the event of a tech failure.
- **Skills gap and training needs**
  Integrating AI into the workplace requires employees to acquire new skills as well as adapt to new technologies, which can be difficult for employees who are not accustomed to using several different technologies. Not all employees are prepared for these technological advancements; therefore, businesses must invest in quality training and development programs to help their workers adjust to these changes and ensure a smooth transition to a technologically advanced environment.

It is important for businesses to consider these advantages and disadvantages before integrating AI into their business practices. While they should consider what works best from a business standpoint, it is also important for them to consider how these technological changes will affect their employees, as well as customers and clients.

AI-DRIVEN SKILLS MENTORING

Badiru (2024) presents a compelling case of the role of mentoring in workforce development. In 2023, the US government announced the infusion of a huge amount of government dollars for the purpose of boosting the productive capacity of the microelectronic industry. The primary purpose is to strengthen the industry. One headline says, "Intel To Receive Up To $19.5B Under CHIPS Act To Build US Semiconductor Manufacturing Facilities." The *chip* on our shoulder stems from the disgust that we had relinquished *chip-making* to foreign nations, both allies and foes. The goal is to bring semiconductor manufacturing back to the USA by pumping billions of dollars into CHIPS Act funding, as the nation ramps up the efforts to lure semiconductor manufacturing back to the USA, where it originally started. The National Science Foundation had advised that breakthroughs in semiconductors and microelectronics would be key to overcoming limits in critical areas, including artificial intelligence, quantum computing, manufacturing, and communications. The huge chips investment by the USA creates a tremendous opportunity for USA manufacturers, but a challenge to develop the required workforce. Workforce development can be boosted by a carefully-structured mentoring program along the spectrum of high school to college. Even reaching further back into grade school is not a bad idea. After all, that younger generation, which is the future workforce, is already conversant with iPads and similar learning devices. Preparing them to learn how to produce the engines that power their iPads will lead to a better and more sustainable appreciation of chip making at the domestic level.

Developing the right workforce for a unique industry never happens by default. A structured program of attracting, cultivating, educating, training, and mentoring must be instituted in order to achieve a sustainable, capable, and reliable workforce for the chips industry. The skills of the workforce must be aligned and integrated into the requirements of the industry. Chip-making is becoming more complex due to advancements in the technology and the process of production. In a systems sense, the preparation of the workforce (the people) must be aligned with what the technology and the process require. As a creative example, the Strategic Ohio Council for Higher Education (SOCHE) has a rigorous program of workforce development for the diverse industries in Ohio, a key part of which is the Defense Industry, the robustness of which affects all other industries. A key part of the SOCHE program is the University-Industry-Government partnership framework. SOCHE partners with academic institutions in a consortium to facilitate workforce-development partnerships. Student internships form the backbone of linking students with industry in some sort of work-learn-for-later-hire relationships.

As the common cliché of "as the military goes, so goes the nation," the defense, aerospace, and aviation sectors have benefited from SOCHE's internship programs. The USA Department of Defense has embarked upon a rigorous program of digital transformation, which further emboldens the need for more chip-making on the domestic front, for national security reasons. One great aspect of the SOCHE approach that may be of interest to business and industry is that the organization embraces the participation of professional organizations in the industry ecosystem. This may encourage professional organizations to bring their systems-thinking "toolbox" to the workforce-development challenge. One systems-based toolbox that SOCHE has embraced is the DEJI Systems Model for workforce-development-program of Design, Evaluation, Justification, and Integration. Working together strategically, we can resolve the chips on our shoulders for targeted microelectronics workforce development. The benefits from that systems-based approach will positively impact other training programs in business and industry.

CONCLUSION

It is safe to say that AI is here to stay. AI can be an extremely useful tool for many different industries as it provides a plethora of solutions to issues that businesses may experience. From automating repetitive tasks, aiding employees in projects by providing quick and accessible information, and enhancing customer and/or client experience, AI can essentially do anything. While this sentiment may scare some people, particularly those worried about losing their jobs to AI, it is important to remember that AI, as it currently is, cannot replace human workers entirely. AI is capable of many things; however, it is not capable of truly understanding human emotion. It merely mimics it, based on what it has been taught. Businesses should keep this in mind when deciding whether or not to incorporate AI into their practices. Overall, AI is an integral part of the workforce and its development. As it continues to evolve each and every day, the workforce will continue to develop, hopefully for the better.

REFERENCES

Badiru, A. B. (June 2024), "CHIP on Our Shoulder Requires Strategy for Workforce Development," ***ISE Magazine***, p. 24.

Byrne, E. (28 August 2024), "Ai in the Workplace: 12 Ways Workplace AI Can Improve the Office." ***OfficeSpace***, www.officespacesoftware.com/blog/ai-in-the-workplace-ai/.

Choudhary, S. (13 August 2024), "Council Post: AI and the Future of Work." ***Forbes***, 2024, www.forbes.com/councils/forbestechcouncil/2024/07/12/ai-and-the-future-of-work/.

Ellingrud, K., et al. (26 July 2023), "Generative AI and the Future of Work in America." McKinsey & Company, www.mckinsey.com/mgi/our-research/generative-ai-and-the-future-of-work-in-america.

Hull, G. (17 July 2024), "How AI Is Already Advancing Retail Training and What's Next." www.retailcustomerexperience.com/blogs/how-ai-is-already-advancing-retail-training-and-whats-next/#:~:text=Leaders%20can%20leverage%20cutting%2Dedge,and%20increased%20customer%20service%20ratings.

Isakova, T. (10 September 2024), "AI Adoption by Industry: Top Use Cases Uncovered," InData Labs. https://indatalabs.com/blog/ai-adoption-by-industry

Khattar, R. (24 August 2023), "Will AI Benefit or Harm Workers?" Center for American Progress. www.americanprogress.org/article/will-ai-benefit-or-harm-workers/.

Kinder, M., et al. (15 October 2024), "Generative AI, the American Worker, and the Future of Work." ***Brookings***. www.brookings.edu/articles/generative-ai-the-american-worker-and-the-future-of-work/.

Lyon, S. (22 January 2025), "10 Pros and Cons of AI in the Workplace," ***People Insight***, peopleinsight.co.uk/pros-cons-ai-workplace/.

National Fund for Workforce Solutions (13 March 2024), "AI and the Future of Work." www.nationalfund.org/ai-and-the-future-of-work/.

Perryman, D. (23 July 2024), "Embracing AI in the Workplace." ***University Human Resources News***, NC State University. news.hr.ncsu.edu/2024/07/embracing-ai-in-the-workplace/.

Quinstein, M. (1 July 2024), "Forget Forgetting: Ai-Driven Strategies for Learning Retention." Training Industry. trainingindustry.com/articles/artificial-intelligence/forget-forgetting-ai-driven-strategies-for-learning-retention/.

Roth, L. (6 September 2023), "How Ai Will and Is Changing the Field of Workforce Development," ***JobsFirstNYC***. jobsfirstnyc.org/latest/how-ai-will-and-is-changing-the-field-of-workforce-development/.

Team, IBM Data and AI (24 January 2025), "Types of Artificial Intelligence," IBM. www.ibm.com/think/topics/artificial-intelligence-types.

Welocalize (16 October 2024), "Boosting Knowledge Retention With AI-Enhanced Training Techniques," Welocalize, Inc. www.welocalize.com/insights/boosting-knowledge-retention-with-ai-enhanced-training-techniques/#:~:text=AI%20excels%20in%20delivering%20real,their%20approach%20and%20progress%20effectively.

What Is the History of Artificial Intelligence (AI)?" (2025) www.tableau.com/data-insights/ai/history. Accessed 25 January 2025.

12 Women in the Workforce

INTRODUCTION

> You need to see something in more than one way to understand it.
>
> **– Ray Solomonoff**

As the opening quote affirms, diversity of thought and views is important in organizational pursuits. Women provide a great platform for seeing things in more than one way. Women already in the workforce and those entering the workforce should be recognized and celebrated for their present and potential contributions on diverse platforms. In times of peace and war, women make tremendous contributions to the society. Women who serve in the military, in particular, bring new enhancements and perspectives to the national mission. This fact is eloquently articulated by Barlow and Norrod (2017). Many times, in post-military engagements, women transition to high-level civilian positions in business, industry, academia, and government.

As of 2021, the world population was estimated to be 7.87 billion, with 50.42% males and 49.58% females. According to the World Bank, the percentage of females in the world population was 49.74% in 2022. Although the share of women in the world varies across countries and regions, and globally in 2020, it generally hovers around 50%. In the United States in 2023, it was 49.8%. Regardless of whichever country or region of interest we consider, that is about half of the population that should be more consistently tapped to be actively and directly involved in the formal workforce for national strategies, goals, and objectives related to national development or sector-based advancement.

As long ago as the 1990s, Badiru (1993) recognized the important roles of women in national industrial development. Women can and are playing exceptional roles in national development through their noticeable participation in the workforce. Women have significant potential in terms of political, economic, and leadership roles in most nations of the world. In developing nations, particularly, women have played important economic roles as traders and leaders for a long time. The essential requirement is a direct and sustained engagement. In very democratic nations, the political power of women has been more pronounced in recent decades. Women's involvement in the workforce should be more aggressively directed at policymaking responsibilities that can facilitate national advancement and stability. Women bring more supportive perspectives to the workforce, including the following characteristics:

- Higher demonstration of empathy
- More sympathetic rationalization
- Better ability to dis-escalate issues
- More neighborly disposition
- More accommodating

DOI: 10.1201/9781003537755-12

- More patient and understanding
- More deliberate in decision making
- More dedicated to missions and goals
- More policy compliant
- More communicative of ideas and reasons
- More amenable to negotiation rather than confrontation
- More rugged assurance of trust
- More creativity
- More adept at improvising
- More listening disposition
- Better ability in handling uncertainty
- More adept at technical knowhow

With all the qualities, characteristics, and attributes, women in the workforce bring more corporate stability, improvement, and advancement to organizations. Statistics over the years bear out these observations.

COLLECTIVE WOMEN EMPOWERMENT

Women groups have emerged in some developing nations to use their collective powers conspicuously in workforce development. For example, the National Council for Women Development has been very active politically in Ghana. In Nigeria, in the 1990s, the Rural Women Development Program, spearheaded by the First Lady of the nation, created programs directed at improving the economic impact of rural women. Small-scale industries owned by women have been one of the tangible outputs of women's collective movements.

On a more recent note in Nigeria, the Lagos State Women Development Center in Agege regularly organizes capacity-building and skill-acquisition seminars for entrepreneurs in the State. The center's programs are aimed at equipping and empowering young female entrepreneurs to learn and develop vocational skills so that they could be self-sufficient, self-reliant, and progressive to become relevant in the economic order of the nation. As the opening quote of this chapter surmises, education empowers. Empowerment is a pathway for women to help change the world positively, in one manifestation of Nelson Mandela's quote that:

> "Education is the most powerful weapon that can be used to change the world."
>
> **Nelson Mandela**

Readers might be interested in the demographical presented in Table 12.1, which focuses on the number of men and women with doctorates in engineering in 2012 in the USA. Although there have been some gains on the women's side, the 2012 data is mostly representative of how much more work the nation needs to do to leverage the available pool of women in the general population for engineering workforce development. The artificial intelligence influences discussed in the preceding chapter opens up new avenues for national considerations. Similar data tabulation can be developed for other educational disciplines, apart from engineering.

TABLE 12.1
2012 Distribution of Men and Women with Doctorates in Engineering

	All	Men	Women	%Women
Total engineering	8427	6527	1883	22.3
Aero/astro engineering	307	268	39	12.7
Chemical engineering	840	583	256	30.5
Civil engineering	496	389	106	21.4
Elec/comp engineering	1943	1598	337	17.3
Industrial engineering	227	165	62	27.3
Materials engineering	743	556	184	24.8
Mechanical engineering	1221	1042	178	14.6
Other engineering	2650	1926	721	27.2

"WOMENTUM"

Women have played and continue to play an active role in society since the beginning or recorded history. To this end, in 2014, as Dean of the Graduate School of Engineering and Management at the USA Air Force Institute of Technology (AFIT), Adedeji Badiru (Badiru, 2014) presented a seminar on "Womentum: The Momentum Created and Sustained by Women in the Intellectual, Technical, Social, Political, Economic, and Family Advancement in the Society." In particular, the seminar highlighted the fact that women are playing diverse roles in many fields. As researchers, educators, technical professionals, scientists, engineers, technologists, mathematicians, mentors, role models, and leaders, women have contributed significant diversity of thought and approach to the advancement of Science, Technology, Engineering, and Mathematics.

WOMEN IN SUSTAINABILITY

Even though women may not generally be the head of the household, they are, nonetheless, leaders of the home. In this leadership role, women can be key players in ensuring sustainability regardless of whichever sustainability definition is in operation. Women play important roles in national development and should lead societal efforts in sustainability. Women have significant capabilities in terms of political, economic, and industrial roles in a nation. In developing nations particularly, women have played important economic roles as traders for a long time. The economic power garnered from their entrepreneurial activities should be transformed into decision-making power to support sustainability programs. In very democratic nations, the political power of women has been more pronounced primarily through voting rights. These rights should be directed at policy-making endeavors that can facilitate sustainability. Three women groups have emerged in some developing nations to better utilize their collective powers. Many of the efforts of these groups have been directed at development. For example, the National Council for Women Development has been very active politically in Ghana. In Nigeria, the Rural Women Development program, spearheaded by the first lady of the nation, is directed at improving the economic impact of rural women. Small-scale industries owned by women have been one of the tangible outputs of these women movements. Other examples abound in developed and developing nations around the world. It is the view of the authors that there should be a global coalition of women for sustainability, whereby an educational platform will be a primary strategy.

THE BETTER HALF

In workforce engagement, women represent the better half, not in the terms of the usual spousal cliché, but in terms of workforce involvement. Women represent half of the U.S. population, but it is still rare for them to have leading roles in setting taxes or budgets in some states, according to a report by Emily Wagster Pettus (Pettus, 2025) on workforce development. Take Mississippi, for example. Only one woman currently serves on the 14-member Joint Legislative Budget Committee. The elite group makes the first recommendations on how much money the state should spend on schools, prisons, Medicaid, and other programs, giving these lawmakers substantial influence over their colleagues and over the lives of people who use government services.

Second-term Sen. Nicole Akins Boyd was appointed to the committee by Lt. Gov. Delbert Hosemann, a fellow Republican who said he did not consider whether to choose a balance of men and women.

> "I don't look at it so much like, 'We need a woman here,' or something like that," Hosemann said. "I look at the abilities and there are plenty of people that have great abilities, male and female. Nicole Boyd, I believe, is outstanding."

Just over 2,400 women are serving in state legislatures across the U.S., or about 33% of the total, according to the Center for American Women and Politics at Rutgers University. Almost 1,600 are Democrats, just over 800 are Republicans and the others are nonpartisan or independent.

Last year, when 31% of state lawmakers nationwide were women, they held about 30% of the seats on committees overseeing taxes, revenues, and other financial matters, according to a review by The Associated Press.

But participation varies widely by state and by session. In Utah, women held only 5% of those seats in 2024. This year, they hold 28%. In Nevada, 62% of finance-related seats were held by women when the Legislature last met in 2023.

Jean Sinzdak, associate director of The Center for American Women and Politics, notes that people who serve in legislatures for a long time tend to receive the most desired committee assignments.

> "Anything budget- and appropriations-related is always one at the top," Sinzdak said. "And so, part of the challenge of getting more women is that women haven't been serving as long and in the numbers needed."

WOMEN ADD TO THE CONVERSATION

Mississippi has the third-lowest percentage of women in its legislature, at 15%, according to the center. The only states behind it are South Carolina, with 13%, and West Virginia, with 11%.

All together, women hold just over 11% of seats on Mississippi's five money committees: Joint Legislative Budget, House and Senate Appropriations, House Ways and Means, and Senate Finance.

Mississippi has never had a woman as governor or House speaker. Only two women have been elected lieutenant governor decades ago. All of those roles are crucial to setting taxes and budgets.

> "Look, I want to see more women there because I think we add to the conversation," Boyd said. "We work together differently, and I think those are all good things for the Legislature."

For example, women in the Mississippi Legislature banded together in 2023 to secure money for evidence testing in rape cases when a backlog at the state crime lab was delaying trials.

"Our male colleagues supported us just as strongly," Boyd said.

BEYOND MONEY COMMITTEES

In West Virginia, Republican Delegate Kathie Hess Crouse said she believes women generally get the committee assignments they ask for. The low number of women on tax and budget committees is not only due to the fact that few serve in the legislature, but also because female lawmakers sometimes prioritize other committees, she said.

That's the case for Hess Crouse, who chaired the House Committee on Workforce Development and served on other committees focusing on energy, manufacturing, and government organization last year. She said legislative leadership has conversations with all the GOP lawmakers before the session starts and asks what their interests are. She always has received her top picks; it just so happens, finance is not one of them.

> "Do I like money? Yes," Crouse said. "I do my home budget. I don't necessarily like it, but I do it. I file my own taxes normally. I work on those things, but it's not my main interest area. So, do I want to serve on finance? No, I have interests elsewhere."

WOMEN IN THE LEAD

The balance is different in Nevada, which in 2019 became the first state where a majority of legislators were women.

That is reflected on key committees. Women have most of the seats on the Assembly's Revenue and Ways, and Means committees. And they hold seven of eight seats on the Senate's finance committee, including the entire Democratic delegation on the committee.

The chair, Sen. Marilyn Dondero Loop, said there could be areas, such as breast cancer funding, where female lawmakers might be more likely to put taxpayer money because of their personal connections to the issue. But she said she doesn't approach her work thinking about gender.

> "Whenever I vote and my other colleagues vote," Dondero Loop said, "we do it solely as being a Nevadan and making things better."

Elizabeth Steiner served as co-chair of the Oregon Legislature's Ways and Means Committee, which handles money matters and has members from the House and Senate, before being elected state treasurer last year. She said it's important for women's personal and professional life experiences to be brought into state decision-making.

> "If you don't include 50% of the population, the perspective of 50% of the population in your decision-making, then you're really disadvantaging everybody: men and women, and certainly children."
>
> **Senator Elizabeth Steiner**

Other states also have women in prominent roles. In Alaska, a Republican woman was one of three co-chairs of the House Finance Committee last year. Connecticut has two Democratic women leading the Appropriations Committee and another co-chairing the Finance Committee. Vermont had two Democratic women in charge of the Appropriations committees for 2023–2024.

South Carolina has never had a woman chair a money committee. Three served on the 25-person House Ways and Means Committee in 2024. A Republican woman was rising in the 23-member Senate Finance Committee, but she took a stand against a total abortion ban and lost her primary along with the only two other Republican women in the South Carolina Senate.

Women and Leadership in National Security

We conclude this chapter with the textual content of the 2015 presentation by The Honorable Deborah Lee James, Secretary of the Air Force on the topic of "**Women and Leadership in National Security,**" which has a direct correlation to the far-reaching national security workforce development of the United States (James, 2015). To preserve the integrity of the presentation, this chapter echoes the discussions directly.

- It's an honor to speak to you today, not only as a woman, but also as only the second female Service Secretary in American history…the first being Dr. Sheila Widnall, who also served as Secretary of the Air Force.

The Case for Diversity and Inclusion

- March is Women's History Month, and 2015 marks the 35th anniversary of the Women's History Movement and the National Women's History Project.

- But why are initiatives like Women's History Month—really, anything that draws the Nation's attention to its diverse population—important? There's simply no other country in the world as widely diverse as the United States of America— or so deeply committed to being unified.
- And yet – as we've learned over time – being unified does not always mean we're leading the charge on *inclusion*; in ensuring that our Nation values, and thus benefits from, the contributions of *all* its citizens.
- Again using women as the example: When President Jimmy Carter designated "National Women's History Week" back in 1980, he said: "…Men and women have worked together to build this Nation. Too often, the women were unsung and sometimes their contributions went unnoticed. But the achievements, leadership, courage, strength, and love of the women who built America was as vital as that of the men…."
- That was in 1980 … and we're still fleshing this challenge out.
- A 2012 study by the Credit Suisse Research Institute found that companies with female directors outperformed similar-sized businesses without women directors by a whopping 26 percent.
- Similarly, the findings of a 2013 report by Thomson Reuters showed that corporations with all-male boards underperformed in comparison to those with gender-diverse boards.
- And yet even findings like these—*proving* the value of the female perspective to corporate performance—has not yet turned the tide.
- Today, women form 47 percent of the labor force—59 percent of the college-educated, entry-level workforce. Yet females only comprise 14.6 percent of executive officers, 8.1 percent of top earners, and 4.6 percent of Fortune 500 CEOs. Females only make up 9 percent of IT management positions, and account for only 14 percent of senior management positions at Silicon Valley startups.
- CNAS's *Battlefields and Boardrooms* ties such findings in the private sector to similar effects among our warfighters…these are challenges I've personally witnessed during my 30 years in both defense and industry, and which still exist across the civilian and government sectors today.
- It's important to note: progress has been made, but we can do better.
- That's how I feel about the Air Force. Progress has occurred … but we can be better.
- For example, we lead the Department of Defense with the highest percentage of women in the Service, but arrive in the middle of the pack for ethnic diversity.
- Our junior enlisted force is pretty diverse, but as rank goes up, the percentage of females and minorities in senior non-commissioned officer ranks goes down. The same holds true for our officers and civilians. In fact, women attrit at twice the rate of men during mid-career.
- The same is true in specific specialties. For example, while African-Americans comprise almost 6 percent of the officer corps, they make up only 2.3 percent of the pilot career field. Similarly, females make up 20 percent of our officer corps, but only 6.7 percent of pilots.

- There have only been eight African-American 4-stars since our inception in 1947, and although our first, General Chappie James, was the first in the DoD, he wasn't a four-star until 1975.
- What's more, the Air Force has never had a female African-American three-star. Further, we have only had one African-American Chief Master Sergeant of the Air Force.
- And—as CNAS, Credit Suisse, Thompson Reuters, and many other studies confirm—this is not just about how we look. It's about our readiness and capability to perform in an increasingly uncertain geopolitical environment.
- To perform, we need top talent. Today we claim the title "World's Greatest Air Force" ...but to remain so, we must learn to be comprehensively inclusive—throughout our ranks, and throughout our specialties. If we get this right, we will glean significant benefit from the many perspectives of the population we serve.

Not a Quick Fix ... But It Must Start Now

- Tomorrow, General Welsh and I will continue Congressional testimony. We'll talk about our fiscal year '16 budget and some of our key acquisition programs ... programs which will take 10 to 20 years to fully mature.
- The question is: are we spending as much time, resources, and energy thinking about the next generation of Airmen as we are the next generation of aircraft?
- Because just like those key acquisition programs, it takes decades to build an Air Force senior leader. As *Battlefields and Boardrooms* notes, our very best newly commissioned female officers, if they started in 2016, would not be eligible for promotion to Brigadier General until 2039.
- Moreover, our National demographics are changing.
- In 2012, Latino, Asian, mixed race, and African American births constituted the majority of births in the US for the first time in our history. It's not just inevitable that our country will experience a demographic shift—it has already happened.
- The strategic framework we published last year, A Call to the Future, recognizes that the biggest threat facing our Nation in the next three decades is the rapid pace of change ... and part of that is demographics. We will need the institutional ability to adapt to this rapid pace of change and keep up—something we call "strategic agility."
- Diversity and Inclusion help us become more strategically agile. Diversity is having the raw material (different types of people) and inclusion is making it work for us by capitalizing on the strengths of differing opinions and approaches.

Initiatives to Strengthen Diversity (What We are Doing About It)

- The Chief and I are passionate about this, and we mean business. Today we are signing two memoranda to all Airmen help guide our efforts.
- The first memo explains, just as I have, why diversity and inclusion are so important to our force.

- The second announces several initiatives which we hope will attract and retain the largest pool of talent to ensure the best 21st century Air Force on the planet. Several of these initiatives have already started; others are on the near horizon.
- We have a total of nine initiatives, and I'll share these nine with you now.

Career Path Tool (CPT) Transformation:

- As we all know, mentorship is important. This is true in both business and government. Mentorship has been very meaningful to me in my career, and I am a big believer that those of us who are more advanced in our careers need to pay it forward.
- Mentors open doors, can help Airmen maximize their strengths, and play an important role in shaping leaders of character, discipline, and vision.
- Our first initiative is The Air Force Career Path Tool (CPT). This is a web-based enterprise solution that supports the Air Force's mentoring program … and we are going to make it better.
- One of our Airmen, Airman First Class Lane Rabern from Air Force Personnel Command, recently suggested a mentor-matching capability, like "match.com", and a skill finder, like craigslist. He submitted his ideas through the "Airmen Powered by Innovation" website, and we listened.
- As a result, we are reconfiguring CPT to provide the best possible support to Airmen as mentors and mentees, creating strong mentoring partnerships and allowing Airmen to be actively involved in their personal development and career management.
- Our reconfigured tool will allow Airmen to find and request mentors; contain forum and live chat capability; and much more. The new system will be named "MyVector" and will be open to both civilians and military. By improving our ability to mentor, we hope to nurture and sustain an Airman's growth throughout the course of their career.
- Mentorship discussions often raise questions about career field development and promotion—and that's where we aimed our next two initiatives.

Diversity and Inclusion Emphasis for Development Team (DT) Boards:

- Our second initiative is to give specific guidance regarding diversity and inclusion to our career field "Development Teams" to help shape our future force.
- By way of background, each development team has a Chairperson who leads the team, and has overall responsibility to shape the career fields for the future.
- For example, within the "13S" space career field, the development team is co-chaired by two 2-star general officers, and assisted by seven senior officers—Generals and Colonels. The development team meets twice a year to review year groups approaching a significant career milestone. The development team identifies the best talent to be made available for critical positions and opportunities for command, as well as developmental education.
- In short, developmental teams are a big deal.

- General Welsh and I will be communicating Diversity and Inclusion Requirements to each of these Development Team Chairs. Specifically, we will ask them to *conduct analyses* to *address barriers* that prevent any of our Airmen from reaching their highest levels of performance.
- After they meet this summer, the DT Chairs will provide those lessons learned and best practices with us, to be shared across the Air Force, and we will seek to bring down the barriers they identify.

Promotion Board Memorandum of Instruction (MOI):

- Our third initiative also gives guidance, this time to our promotion boards. I always provide specific instructions to military promotion board members for every officer promotion and federal recognition board to ensure only the best qualified officers are selected for promotion or recognition. I have the authority to highlight special areas of concern or focus.
- This year, in addition to seeking officers demonstrating commitment to the welfare of our Airmen and to our core values of *Integrity First*, *Service Before Self*, and *Excellence in All We Do*, I have used my authority to instruct board members to find officers who have demonstrated that they will nurture and lead in a diverse and inclusive Air Force culture.
- By focusing on career development and quality officer promotions, it is my hope that we will help shape the Air Force of the future to be more diverse and inclusive than it is today.

Career Intermission Program (CIP)

- We have one more program that's up and running … a program that I announced one year ago. Our fourth initiative is the Career Intermission Program, also known as "CIP".
- Our analysis revealed that life issues contributed to some of our top performers' decisions to depart the service. We developed CIP to allow top performing Airmen the flexibility to transfer from active duty to the Individual Ready Reserve for one to three years to meet personal or professional needs and alleviate some work-life concerns.
- To repeat—more flexibility, we think, is an important factor in order to keep some great Airmen with the Air Force for the long haul.
- In 2014, the first year of implementation, and across the Total Force, 32 Airmen took advantage of this program. In 2015, the program will again be open—for our Total Force. The Air Force will select up to 20 enlisted and 20 officers from the regular Air Force and career status Active Guard and Reserve for the program.
- Let me share one short story of an Airman who we already selected.
- Captain Shannon Williams is a C-21 pilot and flies with the 458th Airlift Squadron at Scott Air Force Base, Illinois. She flies executive airlift, aeromedical evacuation, and is an instructor pilot.
- She's also a new mom. Last Thursday, she gave birth to a baby girl—Jillian—and both mom and baby are happy and healthy. According to Shannon, if CIP was not available, she would have "tried to separate at the earliest possibility."

- Once again, keep in mind—CIP is open to all, but we hope, over time, it will help slow the attrition of females at mid-career.
- Now let me give you a brief overview of what our future holds.
- I've grouped the next four initiatives chronologically, starting with two that target young men and women at the very beginning of their careers.

Increased Female Officer Applicant Pool

- Our fifth initiative aims to increase the pool of female officer applicants.
- There's no question America is blessed to have some of the best and brightest young people in the world. We're even more fortunate when those talented young people seek to join our Air Force as members of our Officer Corps.
- Yet despite this deep pool of talent, our female officer applicants currently comprise only 25 percent of our applicant pool.
- Therefore, we have set an applicant pool goal of 30 percent for our officer accession sources. We're the only Service doing this right now.
- This goal encourages our accession sources to more aggressively compete for our Nation's top female talent. We will accomplish this by adjusting our recruiting strategies to attract talent. Moreover, we will establish partnerships with organizations that mentor and advocate for young women. Over time, we are optimistic about encouraging the next generation of innovative leaders to apply for our officer corps.

Reserve Officer Training Corps (ROTC) Rated Height Screening Initiative

Before I detail our sixth initiative, I'll ask: have any of you watched our Thunderbirds? I've flown with them myself. And if there's one thing I learned the hard way, it's don't eat a heavy lunch before you fly with them!

- Lieutenant Colonel Nicole Malachowski graduated from the United States Air Force Academy in 1996. Colonel Malachowski was the first female pilot selected to fly with the Thunderbirds, and flew the number three jet during the 2006 and 2007 air show seasons. She's an amazing example of the type of Airman we have in our Air Force – and went on to be a White House Fellow afterwards.
- She's also five feet, four inches tall … that's 64 inches. This may seem insignificant, but let me explain.
- In order to fly in our Air Force, current candidates must meet the standard of a standing height of 64 to 77 inches, and sitting height of 34 to 40 inches. We base this on the most restrictive aircraft cockpits. This standard ensures an individual can be qualified in all of our Air Force aircraft.
- But what if Nicole was a ROTC cadet and only five foot three inches tall? That's less than our minimum. Under today's system, we quite possibly would have lost the opportunity to track her into an aviation career … and as you can imagine, that would have been a loss for our Air Force. Fortunately, there's a waiver process if applicants do not meet height requirements. This process allows for additional measurements in order to determine if an individual can safely operate only specific aircraft.

- Currently, only those at the United States Air Force Academy have consistent access to this waiver process.
- Unfortunately, we discovered a number of our highly talented, motivated ROTC cadets did not have the same access. Here's why:
- We have 145 ROTC detachments across the country, many of which are not located in close proximity to an authorized location where the medical teams can take the measurements.
- This eliminated more than one-third (36.6 percent) of our female cadets and a number of male cadets. This is like Team USA trying to recruit Olympians only west of the Mississippi, when we know there are talented athletes across the entire nation.
- Therefore, our sixth initiative is the ROTC Rated Height Screening Initiative. This establishes additional opportunities for ROTC cadets to obtain the appropriate height waivers to maximize our talent base. Specifically, we are moving the measurement process to ROTC field training to reach the greatest number of cadets, and we estimate approximately 900 women will have the opportunity to compete for the rated career fields over the next five years.
- Our goal is to enable more women like Lieutenant Colonel Malachowski, who was recently selected for promotion to Colonel, to do everything from transporting humanitarian aid to flying fighter missions.

Identify Enlisted Airmen for Officer Training School (OTS):

- Moving on to those who are already serving, we have two more initiatives designed to help retain a diverse and inclusive Total Force.
- The seventh initiative recognizes the extraordinary talent resident within our enlisted corps. I point to our own Vice Chief of Staff, General Larry Spencer, as an example.
- General Spencer enlisted in the Air Force in the 1970s. In 1980, he was a Distinguished Graduate from Officer Training School. Today, he's a 4-star.
- How many more Larry Spencer's are already serving?
- Okay, so there's only ONE Larry Spencer … but there are more than 65,000 enlisted members qualified to apply for Officer Training School. Over 6,000 of those individuals are "Honor Graduates" from Basic Military Training, and many of them have diverse backgrounds.
- I'm convinced we can work closely with leaders at all levels to identify and encourage extremely talented, well-educated, and highly motivated enlisted personnel to take on new challenges within the service.
- Our goal is to identify and encourage enlisted personnel to apply for Officer Training School. We're looking for enlisted members who have demonstrated the ability to nurture and lead in a diverse and inclusive Air Force culture and are eligible to attend.
- We will encourage this diverse talent pool to apply for the more than 500 Officer Training School slots, and we'll provide the tools and opportunities to do so, since we believe in growing talent and diversity from all segments of our force.

Post-Pregnancy Deployment Deferment

- Next, do you all agree with me that frequent deployment can take a toll on the family? In some cases, a significant toll?
- For the Air Force, that toll began in the early 1990s and has not stopped. Given world conditions, it is not likely to stop in the foreseeable future. Sadly, some of our most talented Airmen are choosing to leave the Air Force because they are struggling to balance deployments and family issues, especially soon after childbirth.
- To help alleviate some of this strain, our eighth initiative increases our current six-month Post-Pregnancy Deployment Deferment from six months to twelve. Commanders currently have the ability to increase this deferment on a case-by-case basis.
- Case-by-case is not good enough … we will make it the standard!

Use of Panels in Civilian Hiring

- The ninth and final initiative targets our senior civilians. Just as I explained for our senior military ranks, our senior civilian grades are becoming less diverse over time.
- Today, we do not have a standard for civilian hiring panels. Some commands use a panel, others do not.
- Our senior civilians, just like our military leaders, can benefit from diverse backgrounds, experiences, demographics, and perspectives in order to provide innovative leadership, vision, and execution for our Air Force. In order to increase transparency in the selection process and provide hiring officials broader perspectives and insights on the candidates, we will institute and standardize the use of civilian hiring panels for GS-14, GS-15, and equivalent positions.
- These panels will consist of a diverse group of people and will include civilians with no prior military service. Each panel will have a minimum of three individuals, normally senior to the advertised grade of the individuals they are interviewing.
- By building these panels, we hope to highlight the best candidates in each applicant pool while simultaneously increasing diversity in our higher civilian ranks. We know this approach has proved successful in the civilian sector … in fact it has been recognized as an industry best practice. Diverse hiring panels are better able to find the best talent for a given organization or position.

Women in Service Review/Ongoing Focus

- All of these efforts should help advance the ball on diversity and inclusion, but rest assured, General Welsh and I do not consider these "one-shot and done" efforts.
- In fact, we have several more items we're working on to help guarantee we're able to select from the broadest possible talent base while building the most capable Air Force for the future. We will be rolling out our programs with more specifics in the coming months. With respect to the Women in

Service Review that is underway now, General Welsh and I are very firm: there are seven AFSCs that currently remain closed to women, and our position is they should open up, unless someone can prove they should be closed. When they do open – we'll be prepared to help shape and guide increased diversity in those specialties. So, stay tuned!
- In the meantime, we require persistent focus from leaders at all levels to ensure success. Our focus will help capitalize on untapped talent throughout a career, consistently, and over time. I'm confident that everyone – from our senior leaders in Washington, to our Wing commanders, and down to even our newest Airmen – will be catalysts for these positive changes.

CONCLUSION

- The 2015 National Security Strategy reminds us that "Above all, we will take care of our people. We will recruit and retain the best talent while developing leaders committed to an ethical and expert profession of arms."
- Your Air Force will accelerate efforts to capture the best talent America has to offer while matching the changing landscape of our Nation. By affirming our commitment to diversity and inclusion, we amplify the value of our Airmen and their families.
- And we will vigorously develop and reinforce these attributes, emphasizing our Core Values, the importance of dignity and respect in the workplace, and a true appreciation of the contributions of all Airmen to our mission.

REFERENCES

Badiru, A. B. (1993), **Managing Industrial Development Projects: A Project Management Approach**, Van Nostrand/John Wiley, New York, NY.

Badiru, A. B. (18 July 2014), **Womentum: The Momentum Created and Sustained by Women in the Intellectual, Technical, Social, Political, Economic, and Family Advancement in the Society**," faculty seminar presented, Air Force Institute of Technology (AFIT), Dayton, OH.

Barlow, C., and S. Norrod (2017), **Saluting Our Grandmas: Women of World War II**, Pelican Publishing, Elmwood, LA.

James, D. L. (March 4, 2015), "**Women and Leadership in National Security**," Presentation at the National Museum of the USA Air Force, Dayton, OH.

Pettus, E. W. (2025), "Women represent half of the U.S. population but it's still rare for them to have leading roles in setting taxes or budgets in some states," https://apnews.com/article/women-legislators-female-lawmakers-budget-tax-committee-028ba8f32b2968e24af6fb6038daa965, accessed February 20, 2025.

Index

U

V

W